17

Genetic Algorithms and Simulated Annealing

Edited by **Lawrence Davis**
BBN Laboratories
Cambridge, Massachusetts

Genetic Algorithms and Simulated Annealing

Pitman, London

Morgan Kaufmann Publishers, Inc., Los Altos, California

PITMAN PUBLISHING
128 Long Acre, London WC2E 9AN

First published 1987, reprinted 1988

Available in the Western Hemisphere from
MORGAN KAUFMANN PUBLISHERS, INC.,
2929 Campus Drive, San Mateo, California 94403

ISSN 0268-7526

British Library Cataloguing in Publication Data

Genetic algorithms and simulated annealing.—
(Research notes in artificial intelligence,
ISSN 0268-7526)
1. Artificial intelligence
I. Davis, Lawrence II. Series
006.3 Q335

ISBN 0-273-08771-1

Library of Congress Cataloging in Publication Data

Lawrence, David.
Genetic algorithms and simulated annealing.

(Research notes in artificial intelligence)
Bibliography: p.
Includes index.
1. Combinatorial optimization. 2. Algorithms.
I. Title. II. Series.
QA4025.L393 1987 519 87-21357
ISBN 0-934613-44-3 (U.S.)

Reproduced and printed by photolithography
in Great Britain by Biddles Ltd, Guildford

Contents

Acknowledgements

Given the level of research activity in the fields of genetic algorithms and simulated annealing, the series *Research Notes in Artificial Intelligence* with its commitment to rapid production of its volumes is a particularly fitting vehicle for conveying the research reported here. The editor would like to thank Pitman Publishing, Peter Brown of Pitman Publishing, and N S Sridharan, one of the Main Editors of *Research Notes in Artificial Intelligence*, for their help in the production of this volume.

Rob Preuss of Bolt Beranek and Newman Inc. created the LaTeX format for this book and helped with many problems that arose during the conversion of the papers to it.

Finally, the editor would like to extend his thanks to the anonymous referees whose comments, suggestions and intelligent review of the papers helped greatly in the making of this book.

Lawrence Davis and Martha Steenstrup

Chapter 1
Genetic Algorithms and Simulated Annealing: An Overview

This book contains papers about two types of stochastic search techniques—genetic algorithms and simulated annealing. Both techniques have been applied to problems that are difficult and important (designing semiconductor layouts, controlling factories, and making communication networks cheaper, to name a few). Both are modelled on processes found in nature (natural evolution and thermodynamics). Both are being used in artificial intelligence systems that model aspects of human cognition (signal analysis, language processing, and induction, for example.)

Researchers into the application of genetic algorithms and simulated annealing to these sorts of domains have grown more numerous in the past few years, and both areas of investigation are changing rapidly. The authors of the papers in this book are working at the forefront of this research. In this overview, we sketch the nature of the two algorithms and discuss the relation of these papers to the research that is going on in the two fields.

1.1 Genetic Algorithms

John Holland is the founder of the field of genetic algorithms. With the publication of *Adaptation in Natural and Artificial Systems* (Holland 1975), he integrated and elaborated two themes that had persistently recurred in his research: the ability of simple representations (bit strings) to encode complicated structures, and the power of simple transformations to improve such structures. Holland showed that with the proper control structure, rapid improvements of bit strings could occur under certain transformations, so that a population of bit strings could be made to "evolve" as populations of animals do. Holland's book contained some examples of problem domains in which bit strings could be used to encode solutions to problems, with syntactic operations on those bit strings that could be used to alter and improve coded solutions. Holland described the "genetic algorithm", a control structure with which these representations and operations could be managed in order to evolve bit strings that were well-adapted to the problem to be solved. An important formal result stressed by Holland in Holland 1975 was that even in large and complicated search spaces, given certain conditions on the problem domain, genetic algorithms would tend to converge on solutions that were globally optimal or nearly so.

The metaphor underlying genetic algorithms is that of natural evolution. In evolution, the problem each species faces is one of searching for beneficial adaptations to a complicated and changing environment. The "knowledge" that each species has gained is embodied in the makeup of the chromosomes of its members. The operations that alter this chromosomal makeup are applied when parents reproduce; among them are random mutation, inversion of chromosomal material, and crossover — exchange of chromosomal material between two parents' chromosomes. Random mutation provides background variation and occasionally introduces beneficial material into a species' chromosomes. Inversion alters the location of genes on a chromosome, allowing genes that are coadapted to cluster on a chromosome, increasing their probability of moving together during crossover. Crossover exchanges corresponding genetic material from two parent chromosomes, allowing beneficial genes on different parents to be combined in their offspring.

Crossover is the key to genetic algorithms' power. Without crossover, for an individual to acquire two beneficial and unlikely mutations, one of the unlikely mutations must occur by chance to a parent, and then the second unlikely mutation must occur by chance to one of that parent's offspring. It is quite likely that a species reproducing without crossover could have a population containing successful members with one or the other of the two mutations, while no member had both. With crossover, beneficial mutations on two parents can be combined immediately when they reproduce; if the most successful parents reproduce more often than less successful parents and crossover occurs, the probability becomes high that this will happen. This feature of natural evolution — the ability of a population of chromosomes to explore its search space in parallel and combine the best findings through crossover — is exploited when genetic algorithms are used.

A genetic algorithm to solve a problem must have 5 components:

1. a chromosomal representation of solutions to the problem,

2. a way to create an initial population of solutions,

3. an evaluation function that plays the role of the environment, rating solutions in terms of their "fitness",

4. genetic operators that alter the composition of children during reproduction, and

5. values for the parameters that the genetic algorithm uses (population size, probabilities of applying genetic operators, etc.)

A great deal of study has been devoted to each of these five components of genetic algorithms. In the sections that follow, the results of these studies are summarized.

1.1.1 Representation

In all of Holland's work, and in the work of many of his students, chromosomes are bit strings — lists of 0's and 1's. Bit strings have been shown to be capable of usefully encoding a wide variety of information, and they have been shown to be effective representation mechanisms in unexpected domains (function optimization, for example). The properties of bit string representations for genetic algorithms have been extensively studied, and a good deal is known about the genetic operators and parameter values that work well with them.

Some researchers have explored the use of other representations, often in connection with industrial applications of genetic algorithms. Examples of other representations include ordered lists (for bin-packing) embedded lists (for factory scheduling problems), variable-element lists (for semiconductor layout), and the representations used by Glover and by Grefenstette in this volume. These alternative representations tend to be coupled with genetic operators that are different from the ones originally proposed by Holland, and their properties have not been as well studied.

1.1.2 Initialization

Initialization routines vary. For research purposes, a good deal can be learned by initializing a population randomly. Moving from a randomly-created population to a well-adapted population is a good test of the algorithm, since the critical features of the final solution will have been produced by the search and recombination mechanisms of the algorithm, rather than the initialization procedures. For industrial applications, it may be expedient to initialize with more directed methods. Some techniques that have been used include perturbations of the output of a greedy algorithm, weighted random initialization, and initialization by perturbing the results of a human solution to the problem.

1.1.3 Evaluation Function

There are a great many properties of evaluation functions that enhance or hinder a genetic algorithm's performance. One is the normalization process used. When one uses the raw results of a clustering evaluation, for example, a very good individual may score 1020 and a poor one may score 1000. If these scores are used without alteration as measures of each individual's fitness for reproduction, it will take some time before descendants of the good individual outnumber those of the poor one in the population. For this reason, users of genetic algorithms have employed a variety of normalization techniques. The performance of a genetic algorithm is highly sensitive to the normalization technique used, for if it stresses improvements too much it will lead to the driving out of alternative genetic material in the population, and will promote the rapid dominance of a single strain. When this happens crossover becomes of little value, and the algorithm ends up intensively searching the solution space in the region of the last good individual found. If the normalization procedure does not stress good performance, the algorithm may fail to converge on good results in a reasonable time and will be more likely to lose the best members of its population.

Another important feature of the evaluation function is both a strength and weakness of genetic algorithms. Genetic algorithms as they are usually implemented take the single value (or a vector of values) returned by the evaluation function and use that value to determine reproductive fitness. Genetic algorithms are not, in general, sensitive to the way in which evaluation is done. There is no reason they cannot take the workings of the evaluation into account (see Davis and Coombs 1987a). However, dependence on a numeric result of evaluation rather than more complicated feedback has the advantage of making a genetic algorithm robust. When genetic algorithms are applied as optimizers, it often happens that different applications have different requirements. One can incorporate these requirements into the evaluation function without altering the genetic algorithm and genetic operators at all. Systems that are heuristic and finely tuned to their domains may be rendered inoperative when the requirements change. Genetic algorithms are more likely to work well after the evaluation function has been changed in an unexpected way.

An important question to be considered in designing an evaluation function is the implementation of constraints on solutions. Constraints that cannot be violated can be implemented by imposing great penalties on individuals that violate them, by imposing moderate penalties, or by creating decoders of the representation that avoid creating individuals violating the constraint. Each of these solutions has its advantages and disadvantages. If one incorporates a high penalty into the evaluation routine and the domain is one in which production of an individual violating the constraint is likely, one runs the risk of creating a genetic algorithm that spends most of its time evaluating illegal individuals. Further, it can happen that when a legal individual is found, it drives the others out and the population converges on it without finding better individuals, since the likely paths to other legal individuals require the production of illegal individuals as intermediate structures, and the penalties for violating the constraint make it unlikely that such intermediate structures will reproduce. If one imposes moderate penalties, the system may evolve individuals that violate the constraint but are rated better than those that do not because the rest of the evaluation function can be satisfied better by accepting the moderate constraint penalty than by avoiding it. If one builds a "decoder" into the evaluation procedure that intelligently avoids building an illegal individual from the chromosome, the result is frequently computation-intensive to run. Further, not all constraints can be easily implemented in this way.

1.1.4 Genetic Operators

Genetic operators for bit string representations have been extensively studied, while operators for other representation types have not. As one moves from theoretical to applied domains, one tends to use genetic operators taken from and tailored to the domains of application. Candidates for such specialized operators are heuristics that human experts currently use to solve the problem under investigation, and decision procedures that existing algorithms in the domain employ. For examples of genetic operators ingeniously tailored to an application, see Glover's paper in this volume. A good part of the engineering and art involved at present in applying genetic algorithms to industrial problems lies in choosing a chromosomal representation of solutions and a set of genetic operators such that crossover, together with other optimizing operators, moves one rapidly toward promising parts of the search space.

1.1.5 Parameter Values

Effective values of the parameters used in the running of genetic algorithms have been studied intensively for bit string representations, and less intensively for other types of representations. Each combination of genetic operators, representation, and problem has its own characteristics. What ought the probabilities of applying genetic operators in a new domain to be? What population size is best? What combination of operators will work in the new domain? The answers to such questions have been studied, appropriately enough, by the creation of meta-genetic algorithms that take as chromosomes encodings of values for the desired parameters. A meta-chromosome is evaluated by creating a second, object genetic algorithm with the encoded parameter values and running that object algorithm; the evaluation of the meta-chromosome is the evaluation of the best individual found by the object algorithm. Over time, a population of meta-chromosomes tends to converge on coadapted sets of parameter values that out-perform, on average, other sets that have been competing with them. Meta-genetic algorithms of these sorts have been run using bit string representations to encode parameter values and crossover and mutation as genetic operators (Grefenstette 1985). They have also been run using numerical representations with crossover, mutation, and numerical "creeping" operators. Results of both sorts of runs are impressive. The genetic algorithms find better combinations of parameters than those that have been found by the human algorithm designers. Moreover, Grefenstette's combinations of parameter values have been proven useful across a variety of problem domains with bit string chromosomal representations.

1.1.6 Major Areas of Genetic Algorithm Research

There are three major areas in which genetic algorithm researchers are currently working. These areas are: theoretical work in genetic algorithms, applications of genetic algorithms, and classifier systems. These three areas are discussed below, with reference to relevant papers in this volume.

1.1.6.1 Genetic Algorithm Theory

Formal study of the performance of genetic algorithms has been carried out since the field's inception. Two papers in this volume deal with alternatives to the mechanisms that have generally been used. Lashon Booker contrasts several crossover operators and describes the problem types for which each is best suited. David Schaffer analyzes the performance of different procedures used in selecting parents for crossover.

One of the most important questions for genetic algorithm researchers early on was that of determining the characteristics of a problem that make it well-fitted for the genetic approach. Work on this problem by Holland, Bethke, DeJong, and Grefenstette produced a list of features making a domain suitable for the application of genetic algorithms: domains should be multi-modal (or else a hill-climbing algorithm would probably be a better technique); domains should exhibit some, but not a high degree of, epistasis (epistasis is the inhibition of one part of a solution by the action of another); domains should exhibit detectable regularity; and the chromosomal representation should be capable of encoding the regularity of the domain. In this volume David Goldberg describes the "minimal deceptive problem"— the simplest problem that would be expected to mislead a genetic algorithm given these early results — and then produces empirical and formal proof that the genetic algorithm performs well even when faced with problems of this type.

There have been a number of comparisons of genetic algorithms and other types of algorithms in the areas of function optimization, poker playing, and bin packing, to name a few. Such comparisons continue to be made; one of the most recent and most extensive is that carried out by David Ackley in his dissertation. In that work, Ackley compares simulated annealing, a genetic algorithm of his own design, and other search techniques on a variety of different problems. A summary and discussion of Ackley's dissertation research appears here.

1.1.6.2 Applications of Genetic Algorithms

Genetic optimization is a young field, and industrial applications of its theoretical advances are just beginning to take place. The first account of a genetic algorithm being used in industry to carry out commercial optimization appears to be a recent one (Davis and Coombs 1987a). In that paper, the use of a genetic algorithm to optimize the design of a communication network is described. This genetic algorithm has since been applied to other networks and the results have been successful. Descriptions of the system after further refinement are found elsewhere, and are not included here. There are other projects at present concerned with the application of genetic algorithms to the commercial domain, and the number of reports of this type will undoubtedly increase in the near future.

There have been numerous experimental applications of genetic algorithm theory to applied problems. Some — poker-playing, gas pipeline control, and semiconductor layout — have been mentioned already. This volume contains papers describing the application of genetic algorithms to two new domains. Robert Axelrod has continued the research reported in his acclaimed book *The Evolution of Cooperation.* In a paper here, Axelrod reports on the use of a genetic algorithm to evolve successful strategies for playing a game based on the Prisoner's Dilemma of game theory. Axelrod discovered some interesting facts about the domain by observing the evolution of game strategies using genetic algorithms, and he suggests that genetic algorithms may profitably be used to investigate the evolution of other sorts of behavior. David Glover's paper in this volume contains a report of his research into the use of genetic algorithms to evolve efficient mappings of typewriter keys onto character components for languages with large numbers of characters. In Glover's paper use of an ingenious chromosomal representation and some new genetic operators are combined to yield an interesting solution to a complicated problem.

1.1.6.3 Classifier Systems

The principal thrust of John Holland's more recent work has been in the design and use of *classifier systems* – induction systems with a genetic component. With the invention of classifier systems, Holland has provided a framework in which a population of rules encoded as bit strings evolves on the basis of intermittently given stimuli and reinforcement from its environment to "learn" which responses are appropriate when a stimulus is presented. The rules in a classifier system form a population of individuals evolving over time. If the classifier system is well designed, the interactions of the rules and the genetic algorithm will lead to a system that evolves to maximize positive reinforcement from the environment.

In earlier work by Holland and by Wilson, and in work by Booker, classifier systems have been used to model the acquisition of knowledge by entities moving in artificial worlds; other implementations of classifier systems have been concerned with more restricted learning problems. One example of such a system is David Goldberg's dissertation, in which he showed that a classifier system could evolve to control the flow of natural gas through a pipeline, detecting leaks and optimizing profit in the face of seasonal and daily fluctuations in demand.

Stewart Wilson's paper in this volume presents a top-level description of classifier systems. The most extensive description is contained in Holland *et al* 1986. Classifier systems are not described at length in this introduction, because they are composed of much more than a genetic algorithm; the interested reader is referred to Holland *et al* 1986 – a volume with sufficient length to do the topic justice. In their papers here, Holland and Wilson deal with general issues concerning the use a population of classifiers evolving under the control of a genetic algorithm. Holland's paper consists of excerpts from the 1986 volume and commentary on those excerpts. His paper is concerned with showing that the theoretical idea behind his creation of genetic algorithms – that simple structures undergoing simple alterations can evolve to encode tremendously complicated systems – applies in the domain of induction as well as optimization.

Research on classifier systems has concentrated on systems that create their own organization. Large, complicated systems may have difficulty organizing themselves unless they have some organizational primitives available to them when they begin to evolve. Stewart Wilson's paper in this volume describes some organizational primitives and details a technique for executing classifiers based on their position in his organizational hierarchy.

The operations carried out by classifiers – the bit strings that constitute a classifier system's population of rules – are independent and fine-grained. For these reasons, classifier systems are well-suited for implementation on a fine-grained parallel computer. The standard implementation of classifier systems on a sequential computer has been that of Rick Riolo at the University of Michigan. George Robertson's paper in this volume describes some of the algorithmic and implementation details involved in adapting Riolo's system to the Connection Machine of Thinking Machines Incorporated. Robertson's work has shown that classifier systems can be made parallel with near-linear speedup as processors are added.

1.2 Simulated Annealing

Simulated annealing is a stochastic computational technique derived from statistical mechanics for finding near globally-minimum-cost solutions to large optimization problems. Kirkpatrick *et al* (1983) were the first to propose and demonstrate the application of simulation techniques from statistical physics to problems of combinatorial optimization, specifically to the problems of wire routing and component placement in VLSI design.

In general, finding the global minimum value of an objective function with many degrees of freedom subject to conflicting constraints is an NP-complete problem, since the objective function will tend to have many local minima. A procedure for solving hard optimization problems should sample values of the objective function in such a way as to have a high probability of finding a near-optimal solution and should also lend itself to efficient implementation. Over the past few years, simulated annealing has emerged as a viable technique which meets these criteria. Ackley's paper in this volume provides a comprehensive assessment of the relative performance of simulated annealing and other stochastic techniques applied to a variety of optimization problems.

1.2.1 Statistical Mechanics

To appreciate the relationship between techniques in statistical physics and the solution of large optimization problems, one must understand the basics of statistical mechanics.

Statistical mechanics is the study of the behavior of very large systems of interacting components, such as atoms in a fluid, in thermal equilibrium at a finite temperature. Suppose that the configuration of the system is identified with the set of spatial positions of the components. If the system is in thermal equilibrium at a given temperature T, then the probability $\pi_T(s)$ that the system is in a given configuration s depends upon the energy $E(s)$ of the configuration and follows the Boltzmann distribution:

$$\pi_T(s) = \frac{e^{\frac{-E(s)}{kT}}}{\sum_{w \in S} e^{\frac{-E(w)}{kT}}},$$

where k is Boltzmann's constant and S is the set of all possible configurations.

One can simulate the behavior of a system of particles in thermal equilibrium at temperature T using a stochastic relaxation technique developed by Metropolis et al. (1953). Suppose that at time t, the system is in configuration q. A candidate r for the configuration at time $t+1$ is generated randomly. The criterion for selecting or rejecting configuration r depends on the difference between the energies of configurations r and q. Specifically, one computes the ratio p between the probability of being in r and the probability of being in q:

$$p = \frac{\pi_T(r)}{\pi_T(q)} = e^{\frac{-(E(r)-E(q))}{kT}}.$$

If $p > 1$, that is, the energy of r is strictly less than the energy of q, then configuration r is automatically accepted as the new configuration for time $t+1$. If $p \leq 1$, that is, the energy of r is greater than or equal to that of q, then configuration r is accepted as the new configuration with probability p. Thus, configurations of higher energy can be attained. It can be shown that as $t \rightarrow \infty$, the probability that the system is in a given configuration s equals $\pi_T(s)$, regardless of starting configuration, and thus that the distribution of configurations generated converges to the Boltzmann distribution (Geman 1984).

1.2.2 The Annealing Process

In studying such systems of particles, one often seeks to determine the nature of the low-energy states, for example, whether freezing produces crystalline or glassy solids. Very low energy configurations are not common, when considering the set of all configurations. However, at low temperatures they predominate, because of the nature of the Boltzmann distribution. To achieve low-energy configurations, it is not sufficient to simply lower the temperature. One must use an annealing process, where the temperature of the system is elevated, and then gradually lowered, spending enough time at each temperature to reach thermal equilibrium. If insufficient time is spent at each temperature, especially near the freezing point, then the probability of attaining a very low energy configuration is greatly reduced.

The simulation of annealing applied to optimization problems involves the following preparatory steps. First, one must identify the analogues of the physics concepts in the optimization problem itself: the energy function becomes the objective function, the configurations of particles become the configurations of parameter values, finding a low-energy configuration becomes seeking a near-optimal solution, and temperature becomes the control parameter for the process. Second, one must select an annealing schedule consisting of a decreasing set of temperatures together with the amount of time to spend at each temperature. Third, one must have a way of generating and selecting new configurations.

The annealing algorithm proposed by Kirkpatrick consists of running the Metropolis algorithm at each temperature in the annealing schedule for the amount of time prescribed by the schedule, and selecting the final configuration generated as a near-optimal solution. The Metropolis algorithm, which accepts configurations that increase cost as well as those that decrease cost, is the mechanism for avoiding entrapment at a local minimum.

The annealing process is inherently slow. Geman and Geman (1984) determined an annealing schedule sufficient for convergence. Specifically, for a given sequence of temperatures $\{T_t\}$ such that $T_t \rightarrow 0$ as $t \rightarrow \infty$ and $T_t \geq \frac{c}{log\ t}$ for a large constant c, then the probability that the system is in configuration s as $t \rightarrow \infty$ is equal to $\pi_0(s)$. Others have also worked on improving this bound (Hajek 1985, Gidas 1985).

In practice, however, it is often unnecessary to adhere to this conservative schedule in order to achieve acceptable results. Moreover, many applications of simulated annealing map naturally into a parallel processing implementation, and hence the speed of execution can be substantially increased. Nevertheless, determining the proper annealing schedule for a given problem is frequently a matter of trial and error. Davis and Ritter (1987) have used a genetic algorithm to determine better annealing schedules.

1.2.3 Applications of Simulated Annealing

Simulated annealing has been applied successfully to problems in computer design (Kirkpatrick 1983, Vecchi 1983), image restoration and segmentation (Geman 1984, Sontag 1985), combinatorial optimization such as the travelling salesman problem (Kirkpatrick 1984, Bonomi 1984), and artificial intelligence (Hinton 1983).

It is, however, not always trivial to map an optimization problem into the simulated annealing framework. Constructing an objective function that encapsulates the essential properties of the problem and that can be efficiently evaluated, determining a concise description of the parameter configurations and an efficient method for generating configurations, and selecting an effective and efficient annealing schedule all require a certain amount of art.

1.2.4 Boltzmann Machines

The annealing papers in this volume demonstrate novel applications of simulated annealing to traditional problems in artificial intelligence. In each case, the problem is cast as a constraint satisfaction problem represented in terms of a Boltzmann machine – a type of connectionist network in which network states are determined by annealing (Fahlman 1983). The paper by Selman and Hirst demonstrates that parsing can be successfully treated as an optimization problem. The paper by Touretzky and Hinton shows that template matching and variable binding can be viewed as optimization, thus permitting implementation of a production system within the connectionist framework.

The connectionist networks of Boltzmann machines consist of a set of primitive computing elements each possessing a threshold, a binary output, and symmetrically weighted connections between other elements. An element whose output is 1 is said to be "active". The "energy" of a network in a given state is equal to the sum of the thresholds for active elements minus the sum of the weights of connections between active pairs. The output of each element depends upon its "energy gap" – the difference between its threshold and the sum of the products of weight and activity for each incoming connection – which in turn depends upon the state of the network.

The annealing process for Boltzmann machines utilizes a variant of the Metropolis algorithm in which, at a given temperature T, the output of a given element is determined according to its probability

$$\frac{1}{1+e^{\frac{-\Delta E(s)}{T}}},$$

of being active, given that the network is in state s and the energy gap of the element is $\Delta E(s)$.

The weights of the connections between elements actually determine the shape of the "energy" function. A Boltzmann machine can learn the set of weights which yield a desired energy function, simply by obtaining feedback concerning the goodness of the result achieved after each round of annealing and by applying this feedback to a weight-updating mechanism (Hinton 1983).

1.3 Concluding Remarks

The technologies of genetic algorithms and simulated annealing are developing rapidly and are being applied to a wide variety of interesting problems, both inside and outside the domain of artificial intelligence. We hope this volume conveys to the reader a sense of this variety and a sense for the fascinating ways in which these technologies are developing.

David E. Glover

Chapter 2
Solving a Complex Keyboard Configuration Problem Through Generalized Adaptive Search

Abstract

Artificial Intelligence (AI) expert systems for design problems are typically based on heuristic rule sets which guide the design process. Some design/configuration problems, however, are not readily approached by standard knowledge engineering techniques. For instance, it is difficult to apply knowledge engineering practices to large complex combinatorial problems. Typically there are no human experts who solve such problems, logical heuristic characterizations of solution techniques may not be feasible, partial solutions may not exist, and the characteristically large search spaces pose formidable obstacles for traditional search techniques. In such cases, the specific rule-oriented heuristic-based intelligence framework typically used in expert systems may need to be partially or completely supplanted by a more general intelligence-bearing scheme.

A prototype system for solving a complex keyboard configuration problem with these characteristics has been developed in this research. The adaptive search system presented is a variation of the genetic algorithm paradigm (Holland 1975) based on a modified representation and recombinational operator scheme. The prototype algorithm efficiently explores the configuration search space, demonstrating the robustness of the genetic algorithm-type paradigm when applied to problems requiring different representations and operators than those used in the standard genetic algorithm. The results suggest that the technique may be successfully applied to other constraint-based design/configuration problems with similar characteristics.

2.1 Introduction

The solution to a *design problem* is based on the configuring of objects to satisfy a predetermined set of constraints (Eastman 1981, Fenves and Norabhoompipat 1978, Freeman and Newell 1979, Waterman 1985). A wide range of AI expert systems for design problems have been developed in recent years to solve problems such as configuring computer systems (McDermott 1980, McDermott 1981), producing integrated circuit layouts (Grinberg 1980, Kim and McDermott 1983), and designing complex organic molecules (Buchanan and Feigenbaum 1978). Other potential application areas include budgeting, architectural building design, and configuring specialized complex keyboards. Work in the related constraint-based area of *task planning* includes solutions to problems such as job shop scheduling (Davis 1985a, Davis 1985b) and gene cloning experiment planning (Stefik 1981a, Stefik 1981b).

Although most expert systems for design are based on the current *knowledge engineering* paradigm, some problems are not easily approached by heuristic rule-oriented approaches for several reasons. The combinatorial complexity of such problems is a major deterrent to the application of most simple solution strategies. There may not be human experts who routinely solve the problem. The formulation of heuristics for stepwise design construction may not be feasible or possible. Partial solution evaluation may not be possible since a design fragment often cannot be evaluated except in the context of the entire design. The characteristically large nonlinear, discontinuous, and multidimensional search spaces associated with such problems cause difficulties for traditional search techniques[1].

One such problem is the complex *keyboard configuration* problem described in section 2.2. An alternative to the standard knowledge engineering approach and traditional AI search techniques based on a variation of the genetic algorithm paradigm (Holland 1975) is presented. These adaptive search strategies provide an alternative method for rapidly developing part or all of an expert system for specially suited design problems. The heuristic intelligence in such adaptive systems is embedded in the representation/syntactic-recombination scheme of the algorithm.

The configuration problem is now described, followed by an overview of Holland's genetic algorithm paradigm. The Configuration Adaptive System Algorithm (CASA) prototype is presented next, followed by a discussion of experimental results.

2.2 The Keyboard Configuration Problem

This work was motivated by a configuration problem encountered when linguists attempt to produce language-to-keyboard mappings for East Asian languages[2]. For each language, the basic "words" are pictograms called *ideograms*. Each ideogram (language character) is represented by a sequence of language primitives or *components*. These components are printed on top of keyboard keys, and an appropriate sequence of keystrokes followed by "return" represents each character. Due to the large number of components required in the primitives' "alphabet", multiple (*e.g.*, 2-4) component identifiers must reside on top of each key of the standard ASCII keyboard. Without using a shift-type mechanism to distinguish components on each key, the series of keystrokes representing a language character (terminated by a return) is expected to result in an unambiguous table lookup.

[1] See Stefik et al. 1983b, pp 84-85.

[2] Huang (1985) and Becker (1985) provide an overview of the issues involved in East Asian language input systems, and Cui (1985) investigates the problem of analyzing and comparing Chinese character keyboard systems.

The appropriate character, in ideogram form, then appears on the screen.

A design configuration system was needed to generate keytop configuration arrangements for the entire board which would minimize the number of duplicated *keystroke sequences* (ambiguous table references) representing the original *component sequences* used for the language characters.

Prior to the configuration stage, the linguist designer has (1) determined a suitable set of radical primitives ($\approx$ 160 in number) to generate all characters ($\approx$ 3000-5000 in number) in a specific language, (2) decomposed each character into a sequence of radical components (2-4 components in length)[3], and (3) specified a series of layout and permutation constraints to be followed in placing the components on the keys of the board. Each component belongs to a single *component group.* The components of a component group always appear in a specific position on the keys of a specific *key group* (*e.g.*, upper left corner of keys containing four components per keytop). Each component group appears in only one key group. The resultant keyboard configuration comprises a fixed organization of key groups and component groups, and variable permutations of the components within each component group. Functionally, these constraints aid the keyboard user in the process of learning keystroke sequences, since related components are grouped according to specific key quadrant positions on a particular group of keys (key group).

A simple case of a five-key keyboard and seven sequences will serve as an example. (A full scale problem may involve 45-48 keys and 3000+ sequences). The input specifications and constraints are supplied by the linguist. The five keys are divided into two key groups, where group I has four components per key and group II has two components per key. Component groups 1-4 are associated with key group I, and component groups 5-6 with key group II. Components are labeled by integer identifiers. The group structure is shown below.

key group I: (keys 1-3)	component group 1 (upper left): components #'s 1-3
	component group 2 (upper right): components #'s 4-6
	component group 3 (lower right): components #'s 7-9
	component group 4 (lower left): components #'s 10-12
key group II: (keys 4-5)	component group 5 (left side): components #'s 13-14
	component group 6 (right side): components #'s 15-16 .

The linguist has also provided the component sequence set (input sequence set or ISS) for the ideograms of the language. There are seven characters in this example, and all sequences have a length of two or three primitives.

[3] The component sequences must express the semantic intent found in the classical writing of the ideogram while keeping the number of keystrokes entered to a minimum. They cannot be arbitrarily assigned.

ISS

a.	3	15	
b.	16	15	
c.	15	16	
d.	2	10	9
e.	8	7	11
f.	1	9	15
g.	7	6	6

The rating for a keytop configuration (the configuration map or *CM*) is the number of duplicate sequences found in the key sequence set (the output sequence set or *OSS*) produced by the $ISS \times CM \rightarrow OSS$ mapping. Writing each key's components clockwise from the upper left, we produce this configuration map, CM_1.

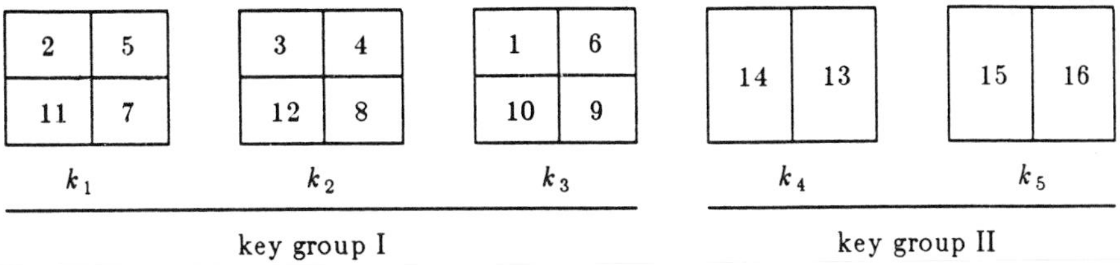

k_1		k_2		k_3		k_4		k_5	
2	5	3	4	1	6	14	13	15	16
11	7	12	8	10	9				

key group I (k_1, k_2, k_3) key group II (k_4, k_5)

This CM produces a two-rating OSS with duplicates [b,c] and [d,g].

OSS (ISS mapped through CM_1)

a.	k_2	k_5	
b.	k_5	k_5	
c.	k_5	k_5	
d.	k_1	k_3	k_3
e.	k_2	k_1	k_1
f.	k_3	k_3	k_5
g.	k_1	k_3	k_3

Note that exchanging the positions of components 2 and 3, and components 14 and 15, would result in a duplicate-free OSS.

The task of the design/configuration system is to search the space of CM's defined by the linguist's constraints to find CM's resulting in near-optimal low-rating OSS's for the input ISS. Finding minimized solutions for non-trivial keyboards is not easy due to the size and complexity of the CM search space (up to 10^{180} points for the full scale problem). A complexity analysis for the keyboard keytop configuration problem has been presented in Glover (1986). It has been shown that in general the density of "good" solution points in the search space may be very low or virtually non-existent. This indicates that the probability of finding zero or near-zero rating solution points may be vanishingly small even if exhaustive search of the space was possible. A powerful

ISS-independent search technique is required to explore these large, multidimensional, nonlinear, discontinuous search space of fixed component permutation groups. In the prototype system investigated in this research, the heuristics used to guide such a search are embedded in the structure of the CM representation and the corresponding syntactic structure recombination operators.

2.3 Holland's Genetic Adaptive Algorithms

One search technique which exhibits the robust quality desired for the CM search space is called the genetic algorithm (GA) developed by John Holland (Holland 1975). Due to representation and constraint mismatches, however, Holland's formulation cannot be directly applied to the keyboard problem. The Configuration Adaptive Search Algorithm (CASA) prototype developed in this research represents a variation of the GA theme which is matched to the problem's constraints. The reader who is familiar with the standard genetic algorithm formulation may exercise the option to proceed directly to section 2.4.

Empirical studies have demonstrated the capabilities of GA's in many diverse areas including function optimization (DeJong 1980), model-fitting (DeJong and Smith 1981), cognitive modeling (Holland and Reitman 1978), meta-level learning (Rendell 1983, Rendell 1985), multi-objective learning (Schaffer 1985), multi-class pattern discrimination (Schaffer and Grefenstette 1985), image registration/pattern recognition (Fitzpatrick et al. 1984, Englander 1985, Wilson 1985), rule discovery (Holland 1980), expert system heuristic rule-set generation (Smith 1983, Smith 1984), pattern tracking models of adaptation (Petit and Swigger 1983), selection of abstract data type implementations (Ford and Norton 1985), bin-packing (Davis 1985b, Smith 1985), compaction of symbolic layouts (Fourman 1985), job shop scheduling (Davis 1985a), the "traveling salesman" problem (Goldberg and Lingle 1985; Grefenstette, Gopal, et al. 1985), and dynamic system control (Goldberg 1985a, Goldberg 1985b).

In essence, genetic algorithms are based on the natural selection principles which guide the generation of new chromosomes in living entities via genetic recombination. In Holland's genetic-based terminology, each gene in the gene pool may take any of several forms called *alleles.* The *genotypes* of the system are then all possible genetic structures that may be created from different combinations of alleles, and a *phenotype* is the totality of the observed behavior of a genotype. A genetic algorithm is composed of a *reproductive plan* which provides an organizational framework for representing the pool of genotypes of a generation, for selecting successful genotypes to be used in creating the offspring of the next generation, and a set of *idealized genetic operators* (i.e., "crossover", "inversion", and "mutation") used to create the new offspring. Holland has shown that the combined effect of the operators results in the propagation of *coadapted sets of alleles,* the building blocks of the selected successful genotypes, generation after generation. The performance of structures in the space focuses the search process in the most promising areas of the search space[4]. The GA paradigm optimizes both global and local search, and the algorithm exhibits "... a synergistic balance in its *exploitation* of knowledge structures that have led to good performance in the past and its *exploration* of new structures in the space ..."[5]

A greatly simplified overview of the reproductive plan and the idealized GA operators follows. Initially, a knowledge base of 50 bit-string GA structures is generated randomly. Each structure is evaluated according to domain specific criteria and assigned a measure of rating or *utility.* Selection probabilities are then computed for each structure based on its utility, with proportionally higher probabilities assigned to higher utility structures. The next generation of structures is then created by selecting structures according to the probabilities and applying the recombination operators described below. Each operator is responsible for creating a fixed percentage of the new generation structures. Current generation structures are used as often as they are selected from the pool. When all fifty new structures have been created, the new generation knowledge base replaces the current (old) generation base. The new structures are then evaluated, and the cycle repeats until a specified number of generations have been completed. The highest-rated structures are copied to an independent data structure as search continues, and are printed at the end of the run.

The three GA recombination operators modify the bit-string structures to create new structures. *Crossover* takes two selected current generation structures, splits the strings at the same randomly determined point, and then creates two new generation structures by swapping the tail portions of the strings. For example, the two structures '10010110' and '11100101', split after the third bit as shown below in (a) and (b), would create structures (c) and (d) after the swap.

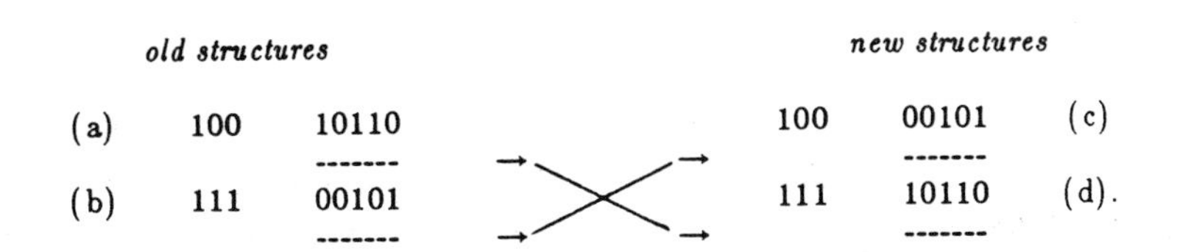

The function of crossover is to generate rearrangements of coadapted groups of bits from high performance structures. It does not create any new "genetic material" in the knowledge base. *Mutation,* on the other hand, randomly changes a bit in a structure, thereby introducing new material into the knowledge base. This operator is assigned a very low percentage of activation, causing it to function as a background operator. The final operator, *inversion,* alters the positional linkages of bits in a structure by inverting

[4]A detailed discussion of genetic encoding and information transmission with respect to adaptive information processing may be found in Sampson 1976, as well in Holland 1975.

[5]Smith 1984, p.2.

the sequence between two randomly assigned points in a single structure. For instance, the structure '110011101', randomly cut after the second and sixth bits, looks like '11 0011 101' and creates a new structure '11 1100 101' when the center substring is inverted. Inversion modifies the configuration layout of schemata in a structure, providing a more fertile ground for recombination by crossover.

A wide variety of implementation-tuning practices have been used in genetic algorithm applications. For instance, the size of the knowledge base may range from 20 to 90 structures, the percentage of new structures created by each operator (usually set at about 60-70% crossover, 30-40% inversion, and mutation one out of every 1000 bits) can vary widely, and several policies for combining the effects of operators have been used. Many such application-dependent issues have been discussed in the literature.

The key feature of GA search is that it is domain-independent. If an application problem solution can be represented as a bit-string vector of features, and syntactic manipulation of that structure by the three operators produces meaningful new structures, the the GA is a powerful tool for searching noisy, discontinuous, multidimensional search spaces. The primary source of power in the genetic algorithm approach lies not in the evaluation of individual structures (search space points), but in the way the algorithm exploits, in an implicitly parallel fashion, the wealth of information found in schemata of high performance structures. High performance schemata become the building blocks for future generations of structures and are propagated in parallel throughout the knowledge base, generation after generation. A formal presentation of the properties and characteristics of GA's may be found in Holland 1975.

2.4 The Configuration Adaptive Search Algorithm (CASA)

Although the robust character of the GA search paradigm is well suited to the demands of the keyboard configuration problem, the bit string representation and idealized operators are not properly matched to the multiple permutation group required constraints. For instance, if three bits are used to represent each component of a simple keyboard of only 40 components, it is easy to show that only one out of every 10^{16} arbitrarily selected 120-bit structures represents a legal configuration map (CM) structure[6]. If breaks are restricted to component boundaries (three bit chunks), the maintenance of component permutation groups is not possible under crossover. If crossover cut points are restricted to component group boundaries, the small number of possible cut points (*e.g.*, 3-6) essentially nullifies the operator's recombinative power. Even if such representation problems could be overcome, the syntactic manipulations of the idealized genetic operators do not support the required identification of component schemata representing high performance structures.

[6]Glover 1986, pp. 68-71.

The Configuration Adaptive Search Algorithm (CASA) program developed in this research maintains the external framework of the original GA reproductive plan paradigm but incorporates a new representation and operator set which function in the same "spirit" as their GA counterparts.

The CASA integer-based, multiple-permutation-group representation insures that all representations produced by the operators are in fact legal CM representations. In addition, the representation is syntactically decomposable for the identification of problem schemata. The general CM structure representation, based on component permutation groups (C_i's), keys (k_{ij}'s), and key groups (K_i's), is shown below:

← comp. group C_i →		← comp. group Cj →
← comp. group C_{i+1} →		← comp. group C_{j+1} →
...	...	...
← comp. group C_{i+q_1} →		← comp. group C_{j+q_p} →
k_{11} k_{12} k_{13} ... k_{1m}		k_{p1} k_{p2} k_{p3} ... k_{pn}
K_1: key group 1		K_p: key group p

In this representation, the sample configuration map CM_1, shown earlier, would look like the following, where key groups are partitioned with double vertical lines, and a key's components are listed in clockwise order above it:

<table>
<tr><td>2</td><td>3</td><td>1</td><td rowspan="2">14</td><td rowspan="2">15</td></tr>
<tr><td>5</td><td>4</td><td>6</td></tr>
<tr><td>7</td><td>8</td><td>10</td><td rowspan="2">13</td><td rowspan="2">16</td></tr>
<tr><td>11</td><td>12</td><td>9</td></tr>
<tr><td>k_1</td><td>k_2</td><td>k_3</td><td>k_4</td><td>k_5</td></tr>
</table>

CM_1

GA theory uses the concept of a search space partition or *hyperplane* containing a coadapted bit group (*schema*) contributing to a structure's performance. A specific hyperplane of the search space is expressed in the form '10##0##...#', where '1' in the first position indicates the feature is present, '0' in the second and fifth positions indicate those features are absent, and the '#' indicates a "don't care" condition for the remaining features. An 'x' bit binary knowledge base structure contains $2^x - 1$ search space hyperplanes. Schemata contained within high performance structures represent regions of the search space which contribute to the structure's global rating. Over time, these schemata are propagated throughout the knowledge base and focus the search.

The hyperplanes of the CM search space used in CASA take on a corresponding but different form. Rather than using binary partitions, CM hyperplanes are composed of basic key building block arrangements which are extracted by the operators' syntactic manipulations. For example, any four-component key (such as k_1 shown in section 2.2) has 12 such building blocks: any two component positions, any three components' positions, all four components' positions, or a "don't care" condition for the entire key. Schemata are composed of one of these positional or structural building blocks for each key, and a search space hyperplane is represented as a vector describing the "state" of each key of the CM configuration. In the following sample hyperplane, cells in the CM marked with the 'Ξ' symbol are considered significant, the remainder are "don't cares".

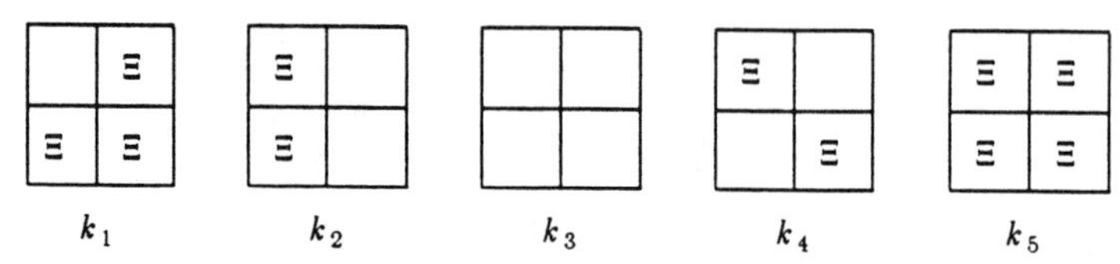

Each component position in the CM can take on any one of the integer values of its C_i permutation group. Whereas the basic hyperplane structural configuration indicates only the component positions of the CM which are significant for the hyperplane, the *dimensions* of a hyperplane express specific values for each component position. A particular dimension of the above hyperplane, characterized as H_{1D1}: ([#,7,12,20], [3,#,#,18], [#,#,#,#], [5,#,14,#], [1,9,15,17]) is shown below.

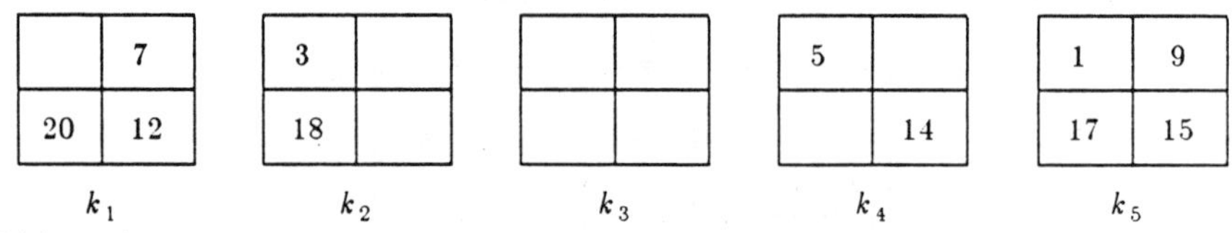

Although the ultimate concern is with the CM structure (the entire configuration) as a whole, the CM's sought are composed of schemata representing high performance hyperplane dimensions of the search space. The approach taken in GA's and in CASA is to allow the reproductive plan via the operators to "break out" and propagate these schemata. An appropriate mapping of the basic schemata building blocks onto the representation and operators which can access these blocks is essential to the plan's success. In CASA, the three idealized GA operators have been replaced by three new adapted genetic operators (AGO's) with five modes of functioning. The AGO's attempt to implement the same qualitative effect on the structures of the CASA knowledge base as the GA operators produce on the bit-string structures. To accomplish this task, the AGO's are designed to function in the integer-based, multiple-permutation group environment of the CM. The AGO's must support the identification and implicit parallel propagation of schemata expressive of high performance hyperplane dimensions of the search space. The three operators presented here are just one possible set which could, in synergistic combination, produce the desired effect. There are two key points with respect to AGO functioning: (1) the operators work with an integer-based rather than binary-based structure, and (2) each operator must maintain the integrity of any C_i component permutation groups that it alters.

2.4.1 The Subgroupswap Operator

The first AGO, corresponding to the GA crossover operator, is the CASA *subgroupswap* operator. Subgroupswap, as the name implies, swaps a portion of the same C_i component group between two structures drawn from the current generation knowledge base to create two new structures for the new generation. In swapping portions of a component permutation group between two structures, there often will be differences in the two subsets of integers swapped. In such a case, the portions of the groups outside of the subgroup swapped will have to be altered to maintain the permutation constraint for the group. Full group swapping (one mode of subgroupswap functioning) is more directly aligned with the spirit of the GA crossover, but the small number of "crossover points" in the CM (*e.g.*, 3-5 points at group boundaries) does not give the operator the recombination power needed. Subgroupswap, therefore, supports both full group and subgroup swapping.

As an example of subgroupswap operation, consider a C_i group, C_1, with component integers 8-14. C_1 is located in a particular key group, and the position of each component indicates its resident key (*e.g.*, the seven positions represent the upper right corner of keys 1-7, respectively). Two permutations of C_1 from two structures $S1$ and $S2$ are shown below:

$S1$: 10 12 8 11 9 13 14
$S2$: 14 11 10 8 12 9 13.

Two cut-points, randomly chosen, define a subgroup sequence. Assume the points are between the first and second components and between the fourth and fifth components. The two groups now have a middle subgroup as shown below:

$S1$: 10   12 8 11   9 13 14
$S2$: 14   11 10 8   12 9 13.

After the swap, the two new structures, $S3$ and $S4$, would look as follows:

$S3$: 12   11 10 8   9 13 14
$S4$: 14   12 8 11   10 9 13.

Since the 8 and 11 are common to both subgroups, making a swap of these elements maintains the integrity of the permutation group. However, moving component 12 from $S1$ into the second position in $S2$ causes a duplicate of that element in $S2$. The conflict is resolved by taking the displaced 11, finding the 12 in the fifth position, and replacing this 12 by the 11. The operator follows the same procedure in moving all elements of the subgroup. Subgroupswap always maintains the integrity of the permutation group, and often this entails disrupting the group beyond the subgroup boundaries.

This operator is similar to the *PMX* operator reported by Goldberg (1985b), where the basic GA paradigm was used to reorder a single route list of integers representing cities for the traveling salesman problem. Goldberg's entire structure is essentially equivalent to one of CASA's C_i groups. Grefenstette, Gopal, et al. (1985) also report the use of GA-based operators for the same problem based on a different "adjacency relationship" scheme where offspring are created by extracting "subtour chunks" from the parents. Other modified versions of GA crossover have also been used by Smith (1985) to solve bin-packing problems where the structure represents a list of boxes to be packed.

The subgroupswap operator is not restricted to swapping a single subgroup. Multiple subgroups of the same key group may be moved in unison between two structures, using the same cut-points in each swap. The number of groups swapped during the operator activation is randomly assigned from 1 to the number of component groups in the key group. To give the operator the greatest degree of generality, subgroupswap treats all groups as rings. If two cut-points in the above example were selected in reverse order, the subgroup excluded by the cuts (*e.g.*, 9 13 14 10 for $S1$) would be used. Use of the "full groupswap" mode causes entire groups rather than subgroups to be swapped.

2.4.2 The Subgroupinvert Operator

The CASA *subgroupinvert* operator is a direct analog of the GA inversion operator. A single structure from the current generation is modified by this operator to create a new structure for the new generation. As in the case of subgroupswap, the group is randomly selected and two randomly determined cut-points mark the beginning and end of the subgroup. The components are then inverted in their subgroup ordering. Inverting the middle segment of $S1$ from above gives the following new ordering:

$S5$: 10 11 8 12 9 13 14 .

If the two cut points had been determined in reverse order (groups are treated as rings), the "external" subgroup would be inverted, resulting in

$S6$: 9 12 8 11 10 14 13 .

Just as in subgroupswap, multiple component groups of a key group (the number randomly assigned) may be inverted in parallel. Whereas the previous operator swapped schemata between two structures, the subgroupinvert operator (1) randomly modifies schemata of the structure and (2) reorders adjacency relationships of components and possibly entire keys. The permutation group constraints are always maintained.

2.4.3 The Componentswap Operator

The final operator, *componentswap*, replaces the GA mutation operator. Componentswap produces a random incremental change in a single structure by swapping two components within a C_i group. This is the smallest atomic change that could be performed on a single structure from the current generation to form a new structure for the new generation. The number of C_i groups, the specific groups, and the elements to be swapped are randomly determined.

In a GA, the mutation operator is used to insure that all points in the search space remain reachable by changing randomly selected bit settings in the structure. This operator functions at a very low probability and, in theory, allows lost allele information to be recovered. Since information cannot be lost in the CM structure, componentswap is used in a higher priority context as an atomic or incremental modifier for a structure. As is the case with the other two operators, the permutation group constraints are always maintained.

The following diagram shows the portion CM (composed of a single key group of n_i keys) altered by two different operations. In the first case, two entire component permutation groups, C_2 and C_4, are swapped in unison with the corresponding C_2 and C_4 groups of another CM structure (not shown) leaving 1/2 of each of the parent structures intact in the offspring. In the second diagram, the identified three subgroups could be swapped in unison with another structure or each subgroup could be inverted within the same structure. It is beyond the scope of this paper to describe the complexity of the potential syntactic recombinations. A detailed analysis of the CM hyperplanes and the effects of the operators on the CM structure and schemata is presented in Glover 1986.

←	group C_1	→
≡	group C_2	≡
←	group C_3	→
←	group C_4	→
k_1	k_2 k_3 ...	k_{n_1}

				≡	≡	≡	≡		
				≡	≡	≡	≡		
				≡	≡	≡	≡		
k_1	k_2	k_3			...				k_{n_1}

The effects of the operators' performance can be summarized as follows:

1. A CM structure may comprise more than one key group. The activation of an operator affects only one key group and a random number of component groups associated with the key group. A multiple key group CM diminishes the portion of the CM and associated schemata changed by an operator.

2. The portion of the CM and associated schemata changed varies widely across the three operators and five modes of functioning, resulting in changes ranging from minute atomic changes (*e.g.*, componentswap) to medium scale swap functions (*e.g.*, subgroupswap, partial groups or two full groups) or random changes (*e.g.*, single full group swap, group or subgroup inverts) to large scale swap or random changes.

3. The operators provide complementary functions. Subgroupswap has the main responsibility of mating high performance structures where the offspring has schemata from both parents. It is the primary operator which drives the search forward based on the structures of previous generations. Subgroupinvert and componentswap are essentially responsible for introducing an element of random change, from a very small to large scale, into the successful CM structures and schemata. The second, and perhaps more important, function of these two operators is to rearrange the adjacency relationships of components, component groups, or keys to facilitate generalized recombination of schemata, particularly for the primary subgroupswap operator.

Choices along the decision tree for the generation of a specific offspring involve the following decision points:

(a) the probability that a given operator/operator mode is selected, which is defined by the operator's activation parameter;

(b) the probability that a given key group is selected;

(c) the uniform random variable determining the number of component groups in the key group to operate on in unison;

(d) the uniform random variable determining the specific component groups in the key group to operate on in unison;

(e) the uniform random variable determining the beginning and ending cut points in groups for non-full group operations;

(f) selection of the parent structure(s) from the current generation to contribute to the generation of the offspring according the parent's selection probabilities.

As noted earlier, the intelligence-bearing heuristics used in CASA are embedded in the structure representation and recombination abilities of the algorithm. This scheme closely models the spirit of the original GA formulation extracted from nature's functional intelligence exhibited in the genetics natural selection processes.

2.5 Experimental Results

The CASA prototype system has been implemented in Berkeley Pascal on a VAX 11/780 operating under UNIX[7]. Test runs have been made on approximately 15 different types of data sets. Each test set includes the input sequence sets (ISS) and the constraints defining the configuration mapping (CM) (see the example in section 2.2). The test sets range from 13-key configurations with 40 components and 300 sequences to 48-key configurations with approximately 160 components and 3150 sequences. The ISS sets used in test runs have sequence lengths of two to four components, and are divided into three categories: (1) randomly generated sets, (2) sets generated from zero-duplicate OSS's, and (3) an actual ISS supplied by a linguist.

The ISS sets contain no duplicates. For category (1), each sequence of the set of size s is randomly assigned a length of 2-4 components, giving s_2 two-component sequences, s_3 three-component sequences, and s_4 four-component sequences. The actual components that appear in each position of each sequence are also randomly chosen from the set of all components in the configuration, regardless of the component group/ key group layouts. Any duplicate sequences are modified so that all sequences are unique. The final step is to further adjust all sequences so that the "key-based constraints" are not violated. This means that for a given key group K_i and sequences of length j, only n_i^j j-component sequences are allowed, and for the total number of keys n, only n^j j-component sequences are allowed. These modifications remove the most gross level of "key-based" errors which would guarantee that no zero-duplicate OSS solutions could exist.

The ISS's in category (2) *guarantee* that at least one zero-duplicate CM solution exists in the search space. This is accomplished by constructing a duplicate-free OSS in the same manner as the ISS in (1), and then inverse-mapping this OSS through a randomly determined CM satisfying all problem constraints to produce a duplicate-free ISS. In this way at least one zero-rating CM is known to exist in the search space, and in general the density of good solutions in the entire population is shifted in the lower direction. Through this technique, the best result found by CASA can be compared to a known best result existing in the CM space.

The single ISS in category (3) is an actual ISS for a real world configuration problem supplied by a linguist. The set comprises two key groups ($n_1 = 35$, $n_2 = 13$), six component groups ($q_1 = 4$, $q_2 = 2$), 160 components, and 3149 sequences ($s = 3149$). This ISS represents the opposite end of the random-ISS spectrum, for it was created by meticulously working back and forth from the ISS to the CM, adjusting first one and then the other in a process of step-wise refinement. Originally this process was done by hand. Later a hill-climbing-based algorithm was used to produce data relevant to fine tuning either the CM or the ISS (within the range of possibilities of the semantic-based constraints) to reduce the number of OSS duplicates. The final result of two years' work

[7]UNIX is a trademark of Bell Laboratories.

by the linguist was a matched CM/KTCS pair resulting in a rating of 183 duplicates in the OSS. Due to the fine tuning the ISS/CM relationship received, it is possible that this test set may have a lower density of good solution points in the space in general. OSS key-based constraints were not observed in constructing the ISS and it is known that no zero-duplicate solutions exist.

In test runs, the percentage of the new generation database created by each operator is fixed by a set of operator parameters. These parameters have been set at (1) subgroupswap (full group mode) - 15%, (2) subgroupswap (regular mode) - 25%, (3) subgroupinvert (full group mode) - 15%, (4) subgroupinvert (regular mode) - 25%, and (5) componentswap - 20%.

Random sampling from the CM search space has been used as a base line measure. Since the distributions produced by random sampling are approximations to the normal distribution, the properties of the standard normal distribution can be used as an indicator of the density of good solution points in the particular search space. The tail probabilities can also be used as an indicator of CASA's efficiency. For instance, in one of the test runs CASA found a best-rated CM solution point of 48 duplicates after 40,550 point evaluations (see Figure 2-1). The probability of a 48-rating solution point *existing* in the search space, according to the random (normal) distribution, is about 10^{-18}. CASA has discovered a point in about 50,000 trials that random search would be expected to turn up (there are no guarantees) about every 10^{18} trials.

Figures 2-1 through 2-4 show the results for four test sets, and are characteristic of all results found. The horizontal axis lists the CM structure ratings found (i.e., the number of duplicates in the OSS produced by the $ISS \times CM \rightarrow OSS$ mapping for the particular CM structure) and the vertical axis gives the number of times structures with these ratings were found. Associated with each graph is the following data: ISS derivation and size; key group constraint information; search space size; for the CASA run: the number of points n in the sample (n = generations $\times 50$), $\overline{x}$ and σ, the low and high rating values found and the point in the search that the low was found; for random search: the number of points n in the sample (n = generations $\times$ 50), $\overline{x}$ and σ, the low and high rating values found, and the probability of finding the best CASA solution by random search. A normal reference curve is provided for each random sample, and another projected normal curve of appropriate scale if the CASA run was for more generations than the random run.

For the 4 examples shown, CASA found low-rating solutions in the evaluation of 13,000–45,000 points that random search would be *expected* to find in 10^{16}–10^{20} samples. The number of generations used in the test runs have been arbitrarily assigned. In most cases, there is no indication that the search is in fact finished. Continuing the runs may produce still better solutions.

2.6 Conclusion

The test results indicate that the adaptive search strategy employed in CASA bears fruit in the search of the complex combinatorial CM search space of the keyboard configuration problem. The natural selection techniques found in genetics display robust search capabilities when applied with representation and operator constraints appropriate to the domain problem, even when these domain characteristics extend far beyond the single bit-string representation scheme developed by Holland. There are a large number of combinatorial design/configuration problems which could be approached by the techniques presented, providing an alternative method for constructing part or all of expert systems for such problems.

There are numerous issues for continued research, including the continuance of the test runs to see if better results are obtained, using different operator parameter settings, and so on. A most interesting analysis would be to compare the results presented here with a simulated annealing program for the same problem run on the same data sets. This would provide an objective measure of CASA's efficiency compared to another search technique known to produce good results in such complex problem domains.

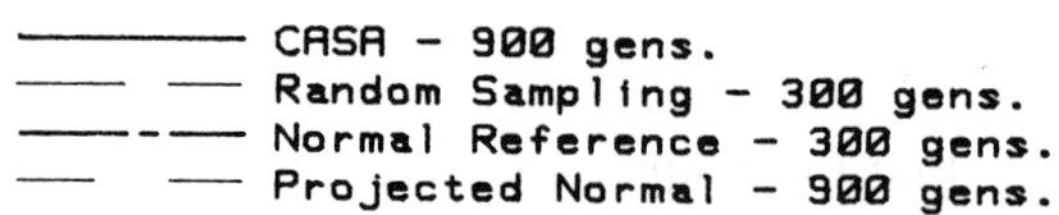

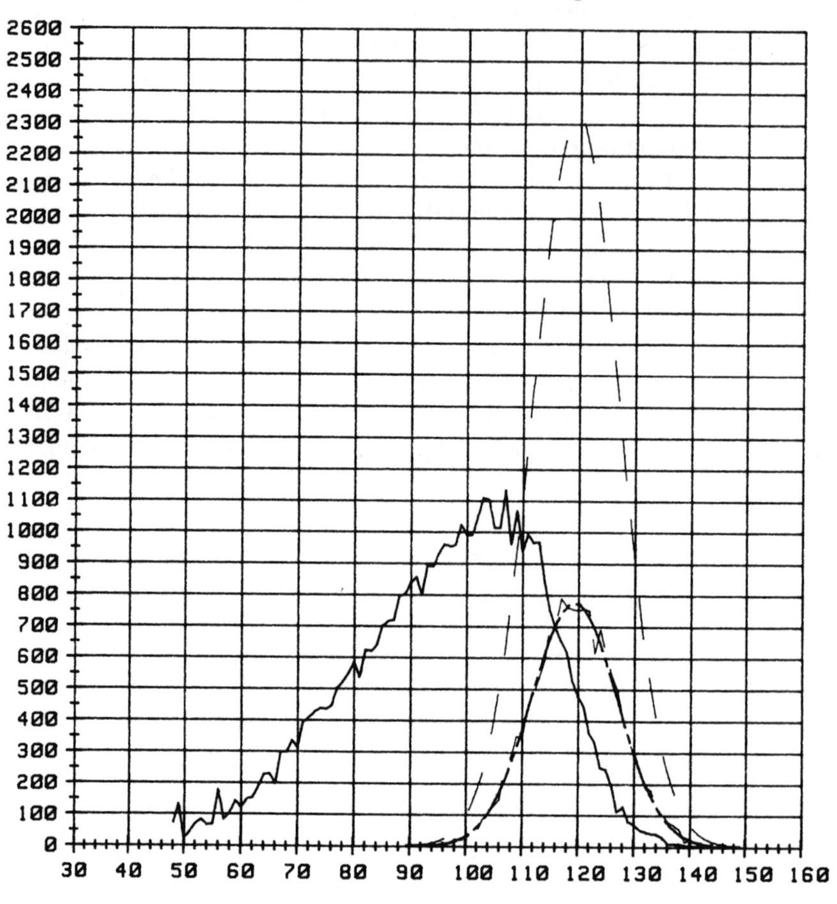

Random ISS - 1500 sequences; 30 keys - 3 groups of 10, each with 4 components per key (120 total components); search space - 10^{79};
CASA: n=45000, $\bar{x}$=96.48, σ=16.89, low=48(40550), high=148
Random: n=15000, $\bar{x}$=119.35, σ=7.70, low=90, high=149,
$P(x \leq 48) = 9.31 \times 10^{-19}$

Figure 2-1

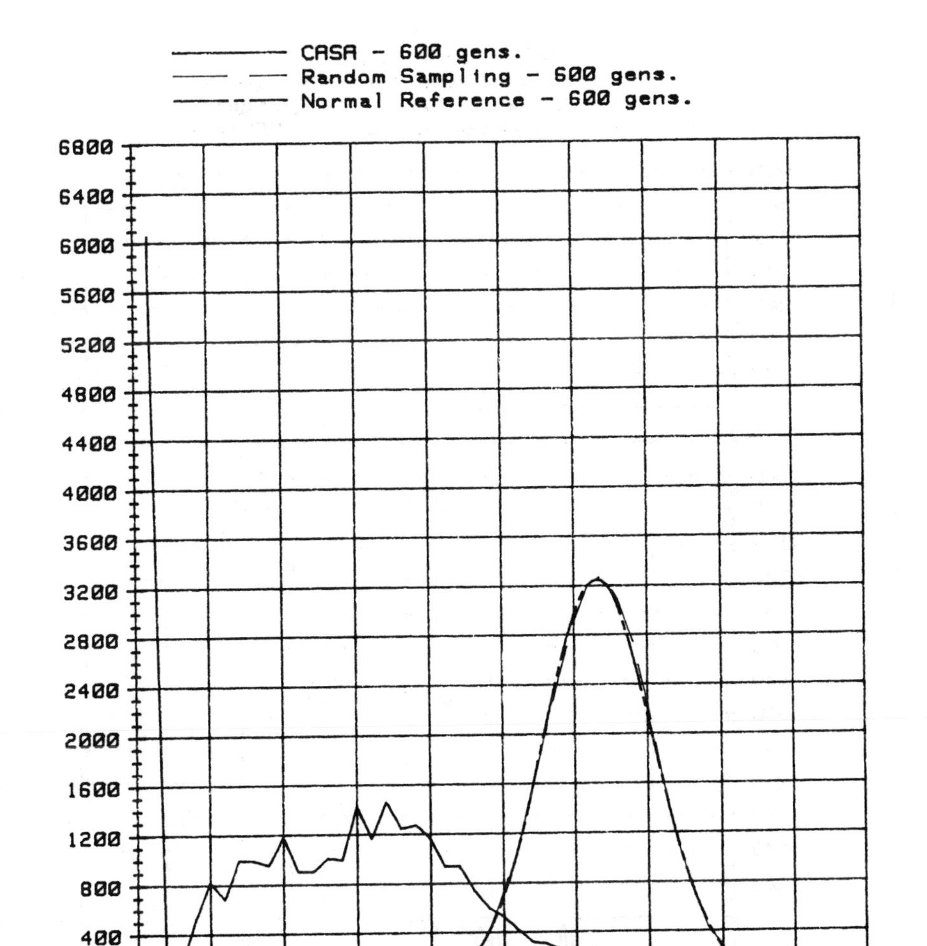

ISS from duplicate-free OSS - 300 sequences; 13 keys - 1 group of 7 (4 components per key), 1 group of 6 (2 components per key) (40 total components); search space - 10^{21};
CASA: n=30001, $\bar{x}$=12.85, σ=8.69, low=1(13250), high=42
Random: n=30000, $\bar{x}$=31.58, σ=3.74, low=16, high=46,
$P(x \leq 1) = 1.31 \times 10^{-16}$

Figure 2-2

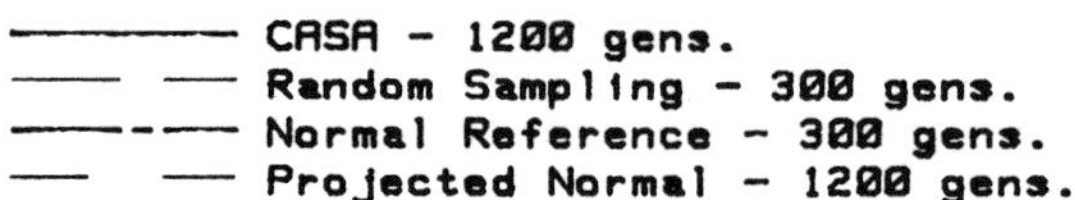

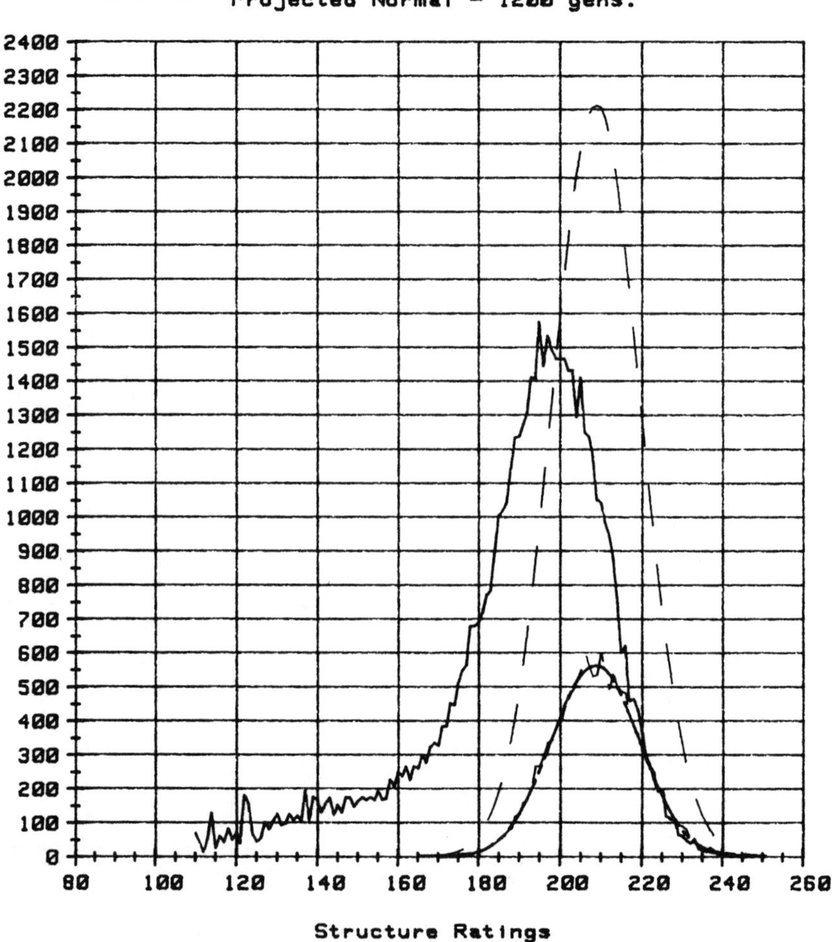

ISS from duplicate-free OSS - 3150 sequences; 48 keys - 1 group of 35 (4 components per key), 1 group of 13 (2 components per key) (160 total components); search space - 10^{180};
CASA: n=60001, $\bar{x}$=190.16, σ=22.18, low=110(55650), high=247
Random: n=15000, $\bar{x}$=208.57, σ=10.79, low=165, high=247,
$P(x \leq 110) = 3.43 \times 10^{-20}$

Figure 2-3

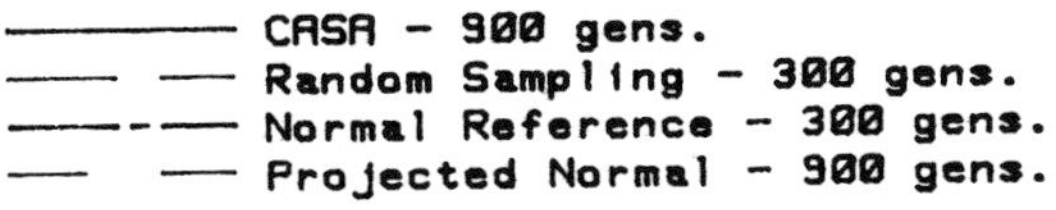

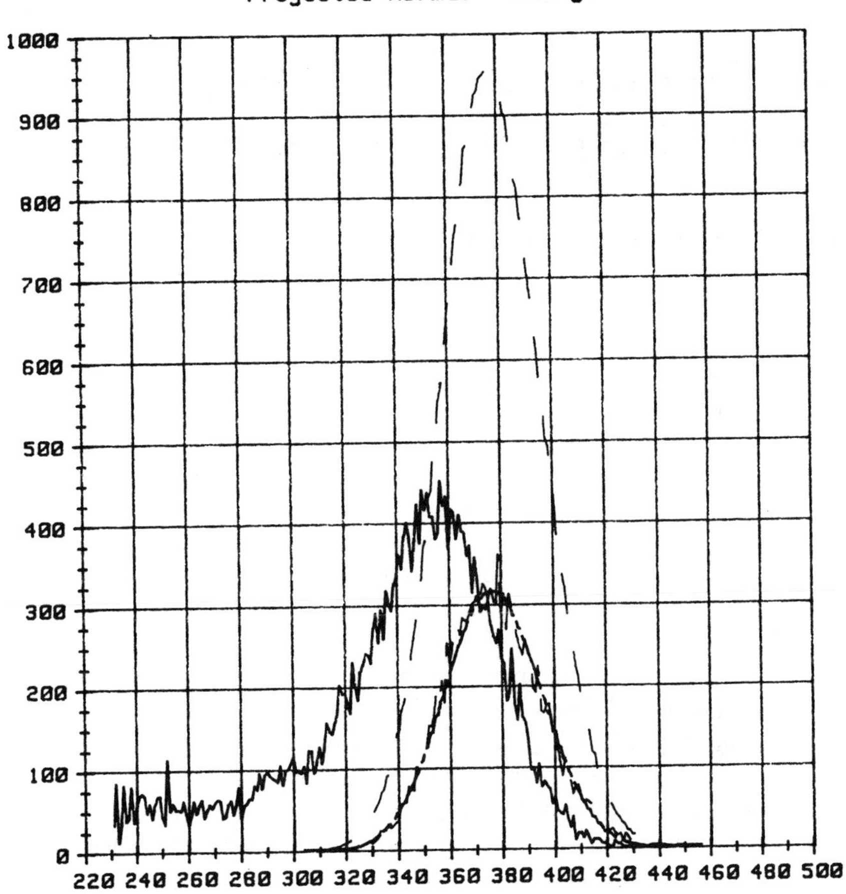

ISS from linguist - 3149 sequences; 48 keys - 1 group of 35 (4 components per key), 1 group of 13 (2 components per key) (160 total components); search space - 10^{180};
CASA: n=30001, $\bar{x}=25.51$, $\sigma=4.50$, low=13(13100), high=40
Random: n=15000, $\bar{x}=376.31$, $\sigma=18.77$, low=304, high=461,
$P(x \leq 231)=4.97 \times 10^{-15}$

Robert Axelrod

Chapter 3
The Evolution of Strategies in the Iterated Prisoner's Dilemma

3.1 The Problem with Cooperation

Mutual cooperation among groups of organisms is a frequently occurring phenomenon. When such cooperation is of benefit to the cooperating agents, and when the lack of it is harmful to them, the mechanisms by which such cooperation might arise and persist seem straightforward.

There are other types of cooperation, however, that are characterized by the fact that, while cooperating agents do well, any one of them would do better by failing to cooperate. For this sort of case, it is more difficult to explain how group cooperation would arise and persist, for one would expect organisms showing a propensity to cooperate to do less well than their neighbors, leading to a dying out of cooperating tendencies in a large, non-cooperating population.

The Prisoner's Dilemma of game theory is an elegant embodiment of this sort of case. In the Prisoner's Dilemma, two individuals can each either cooperate or defect. The payoff to a player affects its reproductive success. No matter what the other does, the selfish choice of defection yields a higher payoff than cooperation. But if both defect, both do worse than if both had cooperated. Figure 3-1 shows the payoff matrix of the Prisoner's Dilemma used in this study.

In many biological settings, the same two individuals may meet more than once. If an individual can recognize a previous interactant and remember some aspects of the prior outcomes, then the strategic situation becomes an iterated Prisoner's Dilemma. In an iterated Prisoner's Dilemma, a strategy is a decision rule which specifies the probability of cooperation or defection as a function of the history of the interaction so far.

To see what type of strategy can thrive in a variegated environment of more or less sophisticated strategies, I conducted a computer tournament for the iterated Prisoner's Dilemma. The strategies were submitted by game theorists in economics, sociology, political science, and mathematics (Axelrod, 1980a). The 14 entries and a totally random strategy were paired with each other in a round robin tournament. Some of the strategies were quite intricate. An example is one which on each move models the behavior of the other player as a Markov process, and then uses Bayesian inference to select what seems the best choice for the long run. However, the result of the tournament

		Column Player	
		Cooperate	*Defect*
Row Player	Cooperate	$R = 3$, $R = 3$ Reward for mutual cooperation	$S = 0$, $T = 5$ Sucker's payoff, and temptation to defect
	Defect	$T = 5$, $S = 0$ Temptation to defect and sucker's payoff	$P = 1$, $P = 1$ Punishment for mutual defection

Note: The payoffs to the row chooser are listed first.

Figure 3-1: The Prisoner's Dilemma

was that the highest average score was attained by the simplest of all strategies, TIT FOR TAT. This strategy is simply one of cooperating on the first move and then doing whatever the other player did on the preceding move. TIT FOR TAT is a strategy of cooperation based upon reciprocity.

The results of the first round were circulated and entries for a second round were solicited. This time there were 62 entries from six countries (Axelrod, 1980b). Most of the contestants were computer hobbyists, but there were also professors of evolutionary biology, physics, and computer science, as well as the five disciplines represented in the first round. TIT FOR TAT was again submitted by the winner of the first round, Anatol Rapoport. It won again.

The second round of the computer tournament provides a rich environment in which to test the evolution of behavior. It turns out that just eight of the entries can be used to account for how well a given rule did with the entire set. These eight rules can be thought of as representatives of the full set in the sense that the scores a given rule gets with them can be used to predict the average score the rule gets over the full set. In fact, 98% of the variance in the tournament scores is explained by knowing a rule's performance with these eight representatives. So these representative strategies can be used as a complex environment in which to evaluate an evolutionary simulation. That is the environment I used to simulate the evolution of strategies for the iterated Prisoner's Dilemma.

3.2 The Genetic Algorithm

The inspiration for my simulation technique came from an artificial intelligence procedure developed by computer scientist John Holland (Holland 1975, 1980, 1986). Holland's technique is called the genetic algorithm. Using a genetic algorithm, one represents strategies as chromosomes. Each chromosome serves a dual purpose: it provides a representation of what the organism will become, and it also provides the actual material which can be transformed to yield new genetic material for the next generation.

Before going into details, it may help to give a brief overview of how the genetic algorithm works. The first step is to specify a way of representing each allowable strategy as a string of genes on a chromosome which can undergo genetic transformations, such as mutation. Then the initial population is constructed from the allowable set (perhaps by simply picking at random). In each generation, the effectiveness of each individual in the population is determined by running the individual in the current strategic environment. Finally, the relatively successful strategies are used to produce offspring which resemble the parents. Pairs of successful offspring are selected to mate and produce the offspring for the next generation. Each offspring draws part of its genetic material from one parent and part from another. Moreover, completely new material is occasionally introduced through mutation. After many generations of selection for relatively successful strategies, the result might well be a population that is substantially more successful in the given strategic environment than the original population.

To explain how this works, consider the strategies available for playing the iterated Prisoner's Dilemma. In particular, consider the set of strategies that are deterministic and use the outcomes of the three previous moves to make a choice in the current move. Since there are four possible outcomes for each move, there are 4x4x4 = 64 different histories of the three previous moves. Therefore to determine its choice of cooperation or defection, a strategy would only need to determine what to do in each of the situations which could arise. This could be specified by a list of sixty-four C's and D's (C for cooperation and D for defection). For example, one of these sixty-four genes indicates whether the individual cooperates or defects when in a rut of three mutual defections. Other parts of the chromosome would cover all the other situations that could arise.

To get the strategy started at the beginning of the game, it is also necessary to specify its initial premises about the three hypothetical moves which preceded the start of the game. To do this requires six more genes, making a total of seventy loci on the chromosome.[1] This string of seventy C's and D's would specify what the individual would do in every possible circumstance and would therefore completely define a particular strategy. The string of 70 genes would also serve as the individual's chromosome for use in reproduction and mutation.

There is a huge number of strategies which can be represented in this way. In fact, the number is 2^{70}, which is about 10^{21}.[2] An exhaustive search for good strategies in this huge collection of strategies is clearly out of the question. If a computer had examined these strategies at the rate of 100 per second since the beginning of the universe, less than one percent would have been checked by now.

To find effective strategies in such a huge set, a very powerful technique is needed.

[1]The six premise genes encode the presumed C or D choices made by the individual and the other player in each of the three moves before the interaction actually begins.

[2]Some of these chromosomes give rise to equivalent strategies since certain genes might code for histories that could not arise given how loci are set. This does not necessarily make the search process any easier, however.

Holland's "genetic algorithm" is such a technique. Genetic algorithms were originally inspired by biological genetics, but were adapted by Holland to be a general problem-solving technique. In the present context, a genetic algorithm can be regarded as a model of a "minimal genetics" which can be used to explore theoretical aspects of evolution in rich environments. The simulation program works in five stages:

1. An initial population is chosen. In the present context the initial individuals can be represented by random strings of seventy C's and D's.

2. Each individual is run in the current environment to determine its effectiveness. In the present context this means that each individual player uses the strategy defined by its chromosome to play an iterated Prisoner's Dilemma with other strategies, and the individual's score is its average over all the games it plays.[3]

3. The relatively successful individuals are selected to have more offspring. The method used is to give an average individual one mating, and to give two matings to an individual who is one standard deviation more effective than the average. An individual who is one standard deviation below the population average would then get no matings.

4. The successful individuals are then randomly paired off to produce two offspring per mating. For convenience, a constant population size is maintained. The strategy of an offspring is determined from the strategies of the two parents. This is done by using two genetic operators: crossover and mutation.

   a. Crossover is a way of constructing the chromosomes of the two offspring from the chromosomes of two parents. It can be illustrated by an example of two parents, one of whom has seventy C's in its chromosome (indicating that it will cooperate in each possible situation that can arise), and the other of whom has seventy D's in its chromosome (indicating that it will always defect). Crossover selects one or more places to break the parents' chromosomes in order to construct two offspring each of whom has some genetic material from both parents. For example, if a single break occurs after the third gene, then one offspring will have three C's followed by sixty-seven D's, while the other offspring will have three D's followed by sixty-seven C's.

   b. Mutation in the offspring occurs by randomly changing a very small proportion of the C's to D's or vice versa.

5. This gives a new population. This new population will display patterns of behavior that are more like those of the successful individuals of the previous generation, and

[3]The score is actually a weighted average of its scores with the eight representative, the weights having been chosen to give the best representation of the entire set of strategies in the second round of the tournament.

less like those of the unsuccessful ones. With each new generation, the individuals with relatively high scores will be more likely to pass on parts of their strategies, while the relatively unsuccessful individuals will be less likely to have any parts of their strategies passed on.

3.3 Simulation Results

The computer simulations were done using a population size of twenty individuals per generation. Levels of crossover and mutation were chosen averaging one crossover and one-half mutation per chromosome per generation. Each game consisted of 151 moves, the average game length used in the tournament. With each of the twenty individuals meeting eight representatives, this made about 24,000 moves per generation. A run consisted of 50 generations. Forty runs were conducted under identical conditions to allow an assessment of the variability of the results.

The results are quite remarkable: from a strictly random start, the genetic algorithm evolved populations whose median member was just as successful as the best rule in the tournament, TIT FOR TAT. Most of the strategies that evolved in the simulation actually resemble TIT FOR TAT, having many of the properties that make TIT FOR TAT so successful. For example, five behavioral alleles in the chromosomes evolved in the vast majority of the individuals to give them behavioral patterns that were adaptive in this environment and mirrored what TIT FOR TAT would do in similar circumstances. These patterns are:

1. Don't rock the boat: continue to cooperate after three mutual cooperations (which can be abbreviated as C after RRR).

2. Be provocable: defect when the other player defects out of the blue (D after receiving RRS).

3. Accept an apology: continue to cooperate after cooperation has been restored (C after TSR).

4. Forget: cooperate when mutual cooperation has been restored after an exploitation (C after SRR).

5. Accept a rut: defect after three mutual defections (D after PPP).

The evolved rules behave with specific representatives in much the same way as TIT FOR TAT does. They did about as well as TIT FOR TAT did with each of the eight representatives. Just as TIT FOR TAT did, most of the evolved rules did well by achieving almost complete mutual cooperation with seven of the eight representatives. Like TIT FOR TAT, most of the evolved rules do poorly with only one representative, called ADJUSTER, that adjusts its rate of defection to try to exploit the other player. In all, 95% of the time the evolved rules make the same choice as TIT FOR TAT would make in the same situation.

While most of the runs evolve populations whose rules are very similar to TIT FOR TAT, in eleven of the forty runs, the median rule actually does substantially better than TIT FOR TAT.[4] In these eleven runs, the populations evolved strategies that manage to exploit one of the eight representatives at the cost of achieving somewhat less cooperation with two others. But the net effect is a gain in effectiveness.

This is a remarkable achievement because to be able to get this added effectiveness, a rule must be able to do three things. First, it must be able to discriminate between one representative and another based upon only the behavior the other player shows spontaneously or is provoked into showing. Second, it must be able to adjust its own behavior to exploit a representative that is identified as an exploitable player. Third, and perhaps most difficult, it must be able to achieve this discrimination and exploitation without getting into too much trouble with the other representatives. This is something that none of the rules originally submitted to the tournament were able to do.

These very effective rules evolved by breaking the most important advice developed in the computer tournament, namely to be "nice", that is never to be the first to defect. These highly effective rules always defect on the very first move, and sometimes on the second move as well, and use the choices of the other player to discriminate what should be done next. The highly effective rules then had responses that allowed them to "apologize " and get to mutual cooperation with most of the unexploitable representatives, and they had different responses which allowed them to exploit a representative that was exploitable.

While these rules are highly effective, it would not be accurate to say that they are better than TIT FOR TAT. While they are better in the particular environment consisting of fixed proportions of the eight representatives of the second round of the computer tournament, they are probably not very robust in other environments. Moreover, in an ecological simulation these rules would be destroying the basis of their own success, as the exploited representative would become a smaller and smaller part of the environment (Axelrod 1984, pp. 49-52 and 203-5). While the genetic algorithm was sometimes able to evolve rules that are more effective than any entry in the tournament, the algorithm was only able to do so by trying many individuals in many generations against a fixed environment. In sum, the genetic algorithm is very good at what actual evolution does so well: developing highly specialized adaptations to specific environmental settings.

In the evolution of these highly effective strategies, the computer simulation employed sexual reproduction, where two parents contributed genetic material to each offspring. To see what would happen with asexual reproduction, forty additional runs were conducted in which only one parent contributed genetic material to each offspring. In

[4]The criterion for being substantially better than TIT FOR TAT is a median score of 450 points, which compares to TIT FOR TAT's weighted score of 428 with these eight representatives.

these runs, the populations still evolved toward rules that did about as well as TIT FOR TAT in most cases. However, the asexual runs were only half as likely to evolve populations in which the median member was substantially more effective than TIT FOR TAT.[5]

So far, the simulation experiments have dealt with populations evolving in the context of a constant environment. What would happen if the environment is also changing? To examine this situation, another simulation experiment with sexual reproduction was conducted in which the environment consisted of the evolving population itself. In this experiment each individual plays the iterated Prisoner's Dilemma with each other member of the population rather than with the eight representatives. At any given time, the environment can be quite complex. For an individual to do well requires that its strategy achieves a high average effectiveness with the nineteen other strategies that are also present in the population. Thus as the more effective rules have more offspring, the environment itself changes. In this case, adaptation must be done in the face of a moving target. Moreover, the selection process is frequency dependent, meaning that the effectiveness of a strategy depends upon what strategies are being used by the other members of the population.

The results of the ten runs conducted in this manner display a very interesting pattern. From a random start, the population evolves away from whatever cooperation was initially displayed. The less cooperative rules do better than the more cooperative rules because at first there are few other players who are responsive — and when the other player is unresponsive the most effective thing for an individual to do is simply defect. This decreased cooperation in turn causes everyone to get lower scores as mutual defection becomes more and more common. However, after about ten or twenty generations the trend starts to reverse. Some players evolve a pattern of reciprocating what cooperation they find, and these reciprocating players tend to do well because they can do very well with others who reciprocate without being exploited for very long by those who just defect. The average scores of the population then start to increase as cooperation based upon reciprocity becomes better and better established. So the evolving social environment led to a pattern of decreased cooperation and decreased effectiveness, followed by a complete reversal based upon an evolved ability to discriminate between those who will reciprocate cooperation and those who won't. As the reciprocators do well, they spread in the population resulting in more and more cooperation and greater and greater effectiveness.

3.4 Lessons

1. The genetic algorithm is a highly effective method of problem solving. Following

[5]This happened in 5 of the 40 runs with asexual reproduction compared to 11 of the 40 runs with sexual reproduction. This difference is significant at the .05 level using the one tailed chi-squared test.

Quincy Wright (1977, pp. 452-454), the problem for evolution can be conceptualized as a search for relatively high points in a multidimensional field of gene combinations, where height corresponds to fitness. When the field has many local optima, the search becomes quite difficult. When the number of dimensions in the field becomes great, the search is even more difficult. What the computer simulations demonstrate is that the genetic algorithm is a highly efficient method for searching such a complex multidimensional space. The first experiment shows that even with a seventy dimensional field of genes, quite effective strategies can be found within fifty generations. Sometimes the genetic algorithm found combinations of genes that violate the previously accepted mode of operation (not being the first to defect) to achieve even greater effectiveness than had been thought possible.

2. Sexual reproduction does indeed help the search process. This was demonstrated by the much increased chance of achieving highly effective populations in the sexual experiment compared to the asexual experiment. If sexual reproduction comes at the cost of reduced fecundity, it is not clear whether this gain in search efficiency would be worth the cost of fewer offspring. However, one case in which search efficiency can be very important is the escape from rapidly evolving parasites. This has been demonstrated in one and two locus models (Hamilton, 1980) and shown to be relevant in the sexual selection of birds (Hamilton, 1982).

3. Some aspects of evolution are arbitrary. In natural settings, one might observe that a population has little variability in a specific gene. In other words one of the alleles for that gene has become fixed throughout the population. One might be tempted to assume from this that the allele is more adaptive than any alternative allele. However, this may not be the case. The simulation of evolution allows an exploration of this possibility by allowing repetitions of the same conditions to see just how much variability there is in the outcomes. In fact, the simulations show two reasons why convergence in a population may actually be arbitrary.

   a. Genes that do not have much effect on the fitness of the individual may become fixed in a population because they "hitch-hike" on other genes that do (Maynard Smith and Haigh, 1974). For example, in the simulations some sequences of three moves may very rarely occur, so what the corresponding genes dictate in these situations may not matter very much. However, if the entire population are descendants of just a few individuals, then these irrelevant genes may be fixed to the values that their ancestors happened to share. Repeated runs of a simulation allow one to notice that some genes become fixed in one population but not another, or that they become fixed in different ways in different populations.

   b. In some cases, some parts of the chromosome are arbitrary in content, but what is not arbitrary is that they be held constant. By being fixed, other parts of the chromosome can adapt to them. For example, the simulations of the individual chromosomes had six genes devoted to coding for the premises about the three moves that preceded the first move in the game. When the environment was the eight representatives, the populations in different runs of the simulation developed different premises. Within each run, however, the populations were usually very consistent about the premises: the six premise genes had become fixed. Moreover, within each population these genes usually became fixed quite early. It is interesting that different populations evolved quite different premises. What was important for the evolutionary process was to fix the premise about which history is assumed at the start so that the other parts of the chromosome could adapt on the basis of a given premise.

4. There is a tradeoff in evolution between the gains to be made from flexibility and the gains to be made from commitment and specialization. Flexibility might help in the long run, but in an evolutionary system, the individuals also have to survive in the short run if they are to reproduce. This feature of evolution arises at several levels.

   a. As the simulations have shown, the premises became fixed quite early. This meant a commitment to which parts of the chromosome would be consulted in the first few moves, and this in turn meant giving up flexibility as more and more of the chromosome evolved on the basis of what had been fixed. This in turn meant that it would be difficult for a population to switch to a different premise. So flexibility was given up so that the advantages of commitment could be reaped.

   b. There is also a tradeoff between short and long term gains in the way selection was done in the simulation experiments. In any given generation there would typically be some individuals that did much better than the average, and some that did only a little better than the average. In the short run, the way to maximize the expected performance of the next generation would be to have virtually all of the offspring come from the very best individuals in the present generation. But this would imply a rapid reduction in the genetic variability of the population, and a consequent slowing of the evolutionary process later on. If the moderately successful were also given a chance to have some offspring, this would help the long term prospects of the population at the cost of optimizing in the short run. Thus there is an inherent tradeoff between exploitation and exploration, i.e. between exploiting what already works best and exploring possibilities that might eventually evolve into something even better (Holland, 1975, p. 160).

3.5 Conclusions

The genetic simulations provided in this paper are highly abstract systems. The populations are very small, and the number of generations is few. More significantly, the genetic process have only two operators, mutation and crossover, and the sexual reproduction has no sexual differentiation and always had two offspring per mating. These are all highly simplified assumptions, and yet the simulations displayed a remarkable ability to evolve sophisticated adaptive strategies in moderately complex environments.

In the future, more complex and realistic simulations are possible. But the main advantage of simulations can already be glimpsed from these minimal simulation experiments. They provide a different intellectual perspective on evolution. Instead of having to rely only on our observations of real biological systems or our standard mathematical models, we will be able to approach genetics and evolution as a theoretical design problem. We can begin asking about whether parasites are inherent in all complex systems, or are merely the outcome of the way biological systems have happened to evolve. We can begin investigating alternative ways genetics might have evolved and see just which properties of our biological heritage are arbitrary and which are not. Today microbiologists are developing the techniques to alter our genetic heritage. Perhaps now is also the time to think about doing some "as if" experiments to better appreciate the fundamental properties of the genetic system that is the basis of our natural endowment.

3.6 Acknowledgements

I thank Stephanie Forrest and Reiko Tanese for their help with the computer programming, Michael D. Cohen and John Holland for their helpful suggestions, and the Harry Frank Guggenheim Foundation and the National Science Foundation for their financial support to Robert Axelrod.

John J. Grefenstette

Chapter 4
Incorporating Problem Specific Knowledge into Genetic Algorithms

Abstract

Like other weak methods, Genetic Algorithms (GA's) are applicable to a broad range of problems for which very little prior knowledge is available. However, many opportunities exist for incorporating available problem specific heuristics into GA-based systems. This paper explores some of these opportunities in the context of the traveling salesperson problem. In particular, several heuristic methods for population initialization, crossover and mutation are discussed and empirical comparisons are presented.

4.1 Introduction

The term *weak method* refers to a problem solving method that makes few assumptions about the problem domain. Weak methods are therefore applicable to a broad range of problems. However, weak methods are often also *low power* methods, meaning that they are not very efficient. Genetic Algorithms (GA's) therefore represent an apparent anomaly — a *powerful weak method.* The power of a GA comes from the sophisticated exploitation of information derived from a relatively limited search effort, providing only that the search space offers some sort of regularity. GA's are able to perform fairly efficient search even if the available knowledge is limited to evaluation procedure that can measure the quality of any point in the search space (Holland 75). This makes GA's a useful search technique for many problems for which the only alternative is some form of random search. However, additional knowledge about the problem in question is often available. In this paper, we investigate the use of additional sources of problem specific knowledge in genetic search procedures.

Consider the usual outline of a GA shown in Figure 4-1. It is possible to exploit problem specific knowledge in virtually every phase of the GA. First, the initial population $P(0)$ might be chosen heuristically rather than randomly, with the goal of introducing some helpful building blocks into the gene pool. This should be done carefully since GA's are notoriously opportunistic and may quickly converge to a local optimum if the initial population contains a few structures that are far superior to the rest of the population.

Another opportunity to use problem specific knowledge is in the recombination

```
procedure GA;
begin
      initialize population P(0);
      evaluate P(0);
      t = 1;
      repeat
            select P(t) from P(t-1);
            recombine P(t);
            evaluate P(t);
      until (termination condition);
end.
```

Figure 4-1: Outline of a Genetic Algorithm

operators. In many successful GA studies, these operators perform simple syntactic operations such as crossover, mutation and inversion. For many problems, however, it may be possible to define recombination operators that take problem specific knowledge into account when constructing offspring structures from parents. The challenge in exploiting problem specific knowledge in the recombination operators is that such operators must not defeat the implicit parallelism provided by the genetic selection rule.

The evaluation of structures can also utilize problem specific knowledge. For example, in GA-based systems for bin-packing (Davis 1985) and for circuit layout (Fourman 1985), the structures in the population are evaluated by heuristic routines for constructing explicit solutions. In these cases, the GA is searching the space of constraints for the heuristic evaluation routines, rather than the space of explicit solutions to the underlying optimization problem.

Finally, it is widely recognized that GA's are not well suited to performing finely tuned local search. Like natural genetic systems, GA's progress by virtue of changing the distribution of high performance substructures in the overall population; individual structures are not the focus of attention. Once the high performance regions of the search space are identified by a GA, it may be useful to invoke a local search routine to optimize the members of the final population.

The remainder of this paper explores some of these issues in the context of the well known traveling salesperson problem.

4.2 The TSP

The Traveling Salesperson Problem (TSP) is popular in part because it is so easy to state yet is so hard to solve. The statement is: Given a set of N cities, find the shortest tour that visits each city exactly once. The TSP is NP-Hard and therefore probably does not have any polynomial time algorithmic solution (Garey 1979). The TSP has served as a test problem for many heuristic search algorithms, so it is natural to attempt its solution by GA's. However, the choice of an appropriate representation of the TSP suitable for genetic operators is a non-trivial task. Consider the *path* representation in which a tour is represented by a list of cities. If we apply the classical crossover operator to the two tours $(abcde)$ and $(adecb)$ by cutting between between the third and fourth cities, we get the offspring $(abccb)$ and $(adede)$, neither of which are legal tours. In an earlier paper (Grefenstette 1985), we describe the so-called *ordinal* representation of permutations for which the classical crossover operator is guaranteed to yield legal tours. However, the performance of the resulting GA is poor. Furthermore, analysis shows that reason for the poor performance is that the offspring produced by crossover do not inherit important combinations of features (i.e., subtours) from the parents. In the current work, we avoid the consideration of the representation of tours by incorporating additional knowledge into the recombination operator. That is, the versions of crossover we consider construct legal tours based on the edges in the parent tours — as long as the crossover routine knows the representation of tours, it does not matter what the underlying representation is.

The algorithms in this study are compared on a suite of 10 test problems constructed by randomly choosing 100 city locations in a square region of the Euclidean plane. An empirical formula for the expected length L^* of a minimal tour for such TSP problems is:

$$L^* = K\sqrt{NR}$$

where N is the number of cities, R is the area of the region and K is an empirical constant that is approximately 0.765 (Stein 1977). In our experiments, we fix the area R so that $L^* = 100.0$. It should be noted that the above formula gives only an estimate of the expected minimum tour length; the exact minimal tour lengths for our test problems are unknown. Nevertheless, this scheme allows us to average the performance of a given algorithm over the set of test problems to obtain a rough measure of efficiency. All experiments described below consist of running a GA against each of the 10 test problems for 20,000 trials. In all experiments, the population size is 100 and the crossover rate is 0.6, as suggested by previous studies (DeJong 1975, Grefenstette 1986).

There are many plausible performance measures for adaptive search algorithms. The cost of the best tour found after a fixed number of trials is a natural metric. In addition to performing efficient search, an ideal GA should maintain a high degree of diversity within the population. Population diversity is crucial to a GA's ability to continue the fruitful exploration of the search space. For this reason, we measure the *entropy* H in a population of tours as follows: For each city i, we compute a measure of the entropy H_i in the set of edges incident on i in the current population using the formula:

$$H_i = \frac{-\sum_{j=1}^{j=N} \left(\frac{n_{ij}}{2P}\right) \times \log \frac{n_{ij}}{2P}}{\log(N)}$$

where n_{ij} is the number of edges connecting city i and city j in the population, N is the number of cities, and P is population size. Note that there are $2P$ edges incident on i. The denominator is chosen so that $H_i = 1$ if all edges are represented uniformly in the population. The population entropy H is then defined as:

$$H = \frac{\sum_{i=1}^{i=N} H_i}{N}$$

Thus, as the population converges, H approaches 0. If H decreases too rapidly, the algorithm is likely to exhibit a premature loss of diversity that makes further search futile. We consider the trajectories of both the best tour length and the population entropy in comparing GA's in the following sections.

4.3 Choosing the Initial Population

In this section we consider the effects of seeding the initial population for the TSP. In most GA studies to date, the initial population consists of entirely random structures. The usual justification is the desire to measure the performance of GA's under the most challenging circumstances, i.e., complete lack of knowledge about the search space. However, for many problems like the TSP, it is relatively easy to formulate reasonably good initial structures for the first population. The methods considered differ to the extent that they rely on knowledge about the edge costs in the construction of the initial population of tours. As a base point, GA A ignores the cost of edges. Instead, the initial population consists of 100 tours constructed using a quasi-random procedure designed to maximize the entropy in the initial population. That is, tours are constructed at random except that edges not occurring in the population are preferred to edges that are already present. GA B uses a probabilistic greedy heuristic in constructing initial tours: start with a randomly chosen city, add an edge to the nearest one in a random sample of unvisited cities, and repeat until all cities have been visited. With a fixed sample size of, say, 20 cities, this procedure produces a better-than-average tour with time complexity $O(N)$. By repeating the algorithm with different starting cities, we obtain a population of good initial tours. Finally, GA C uses a deterministic greedy algorithm: start with a random city, add an edge to the nearest unvisited city, and repeat until all cities have been visited. The greedy algorithm generally yields a sub-optimal solution to the TSP because while the early edges tend to be short, it may be necessary to include several long edges to complete the tour. Computationally, the greedy algorithm runs in $O(N^2)$

Seeding the Initial Population			
	GA		
Problem	A	B	C
1	105.50	102.13	101.82
2	109.60	109.19	111.69
3	106.55	108.21	106.81
4	104.55	101.33	103.31
5	108.53	105.91	107.71
6	107.77	111.72	109.09
7	107.32	106.28	109.45
8	110.88	103.91	110.17
9	105.00	103.59	105.06
10	108.66	107.48	109.92

Table 4-1: Results with Heuristic Initialization

steps per tour and so it can be expensive for large N. As before, we obtain an initial population of distinct tours by repeating the greedy algorithm with different starting cities. Other than the method of selecting the initial population, all other control parameters (population size, crossover operator, crossover rate, etc.) are held constant in GA's A, B, and C.

The results of using the three initialization methods on the 10 test problems are shown in Table 1. GA B, using the heuristically chosen initial population, appears to perform slightly better than GA A, but the Wilcoxon sign test (Golden 1985) fails to indicate a statistically significant difference between GA A and B and the 0.05 level. However, there is a statistically significant difference between B, which uses the probabilistic greedy algorithm, and C, which uses the deterministic greedy algorithm. The trajectories for the best tour length are illustrated in Figure 4-2, where the results are averaged over the 10 test problems. As expected, GA A exhibits a steady rate of improvement and GA B shows the advantage gained by seeding the initial population. In contrast, GA C produces good results very quickly but also levels out rapidly. The trajectories for the entropy are shown in Figure 4-3, which clearly shows that GA C suffers from a severe allele loss during the initial few generations. GA B shows far less initial allele loss due to its probabilistic initialization procedure. In summary, GA B appears to represent a reasonable trade-off between the requirements of efficient search and high entropy in the initial population.

In other experiments, the percentage of the initial population that was heuristically chosen in GA B and GA C was varied between 100% and 10% with the remaining initial tours constructed as in GA A. Surprisingly, no significant differences were observed at these levels when compared to the 100% initialization. Clearly, the heuristically chosen tours were highly favored by selection during the first few generations and the randomly constructed tours were quickly eliminated from the population. These results support

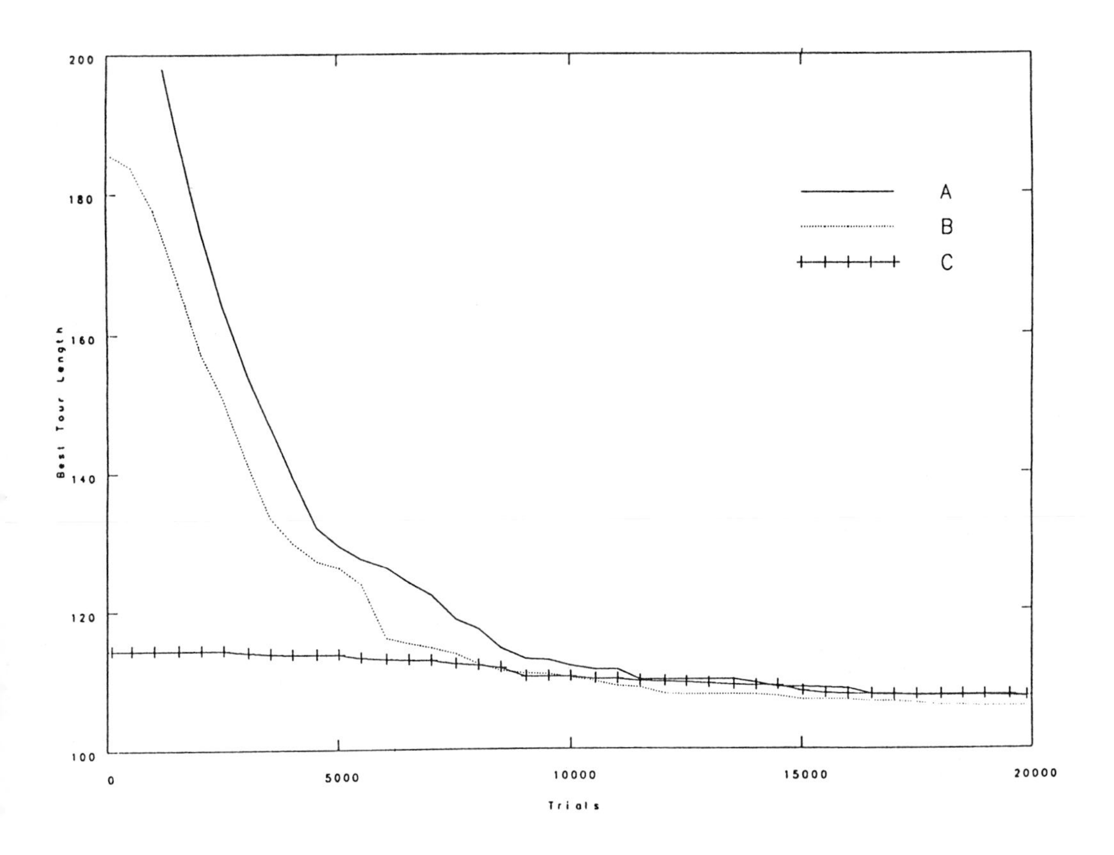

Figure 4-2: Performance of Initialization Methods

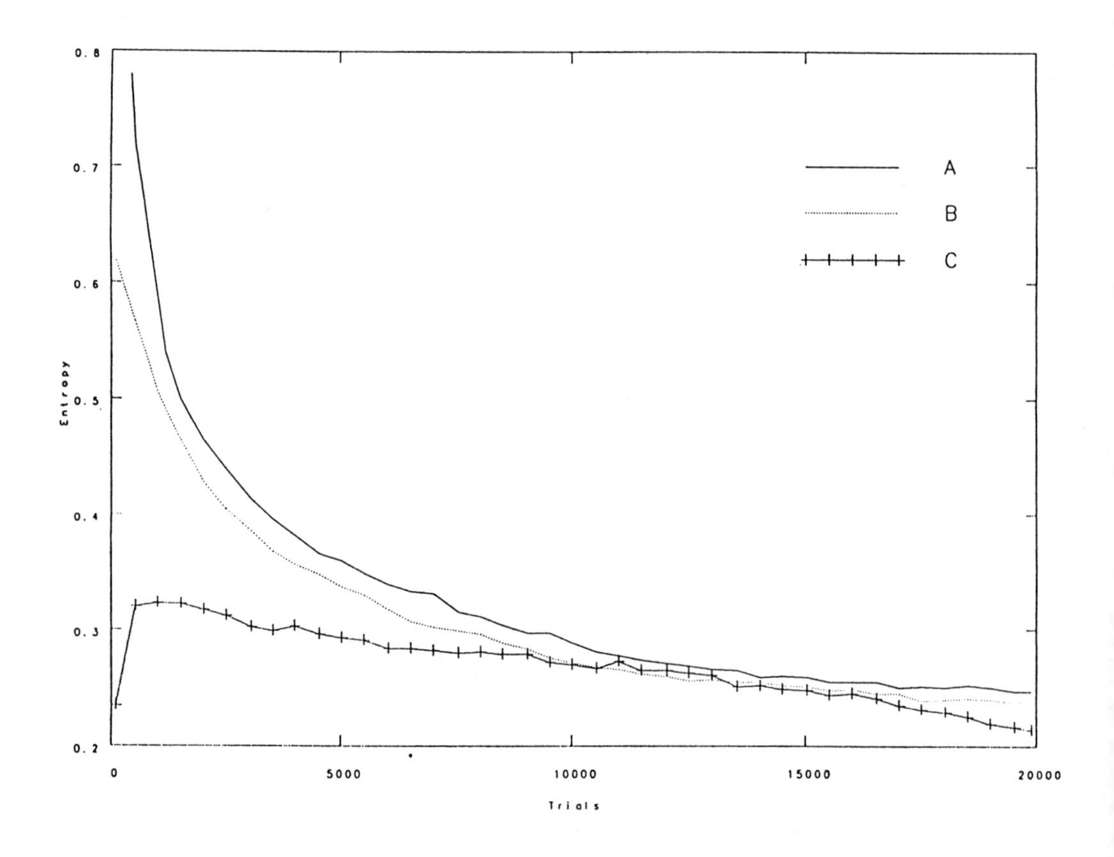

Figure 4-3: Convergence of Initialization Methods

the observation that heuristic seeding of the initial population should be done with care. It is important to maintain the variance of alleles in the initial population in order to provide grist for the genetic mill.

4.4 Recombination Operators

The power of GA's arises from an interaction between the selection and recombination steps of the algorithm. By selecting structures for reproduction such that the number of offspring of a given structure is proportional to that structure's performance (relative to the population average), GA's achieve the important property of *inherent parallelism:* the number of structures that contain any given combination of features changes at a rate roughly proportional to the observed average performance of all structures that contain the given combination of features (Holland 1975). That is, GA's search the space of feature *combinations* by generating and testing points in the space of structures. Since the latter space is exponentially smaller than the former, we obtain a highly efficient search. The role of recombination is to introduce new combinations of features into the population without disrupting the selection process. The crossover operator should favor the inheritance of small building blocks in the expectation that small building blocks associated with high performance will propagate through the population, thereby reducing the dimensionality of the problem and leading to larger and larger inherited building blocks. In this section, we explore the use of heuristic operators that fill the role of the traditional crossover operator.

We first motivate the development of heuristic crossover operators by examining the performance of two "blind" recombination operators that create legal tours without reference to the costs of individual edges. These two crossover operators are described more fully in earlier studies (Goldberg 1985, Grefenstette 1985, Rosmaita 1985), along with many others such as taking alternate edges from each parent or taking subtours of random length from each parent. Let GA D use the following crossover operator:

1. Randomly choose a city as the current city for the offspring tour.
2. Consider the four (undirected) edges incident to the current city in the parents. Define a uniform probability distribution over these edges except that the probability associated with an edge incident to a previously visited city is 0.
3. Select an edge based on this distribution. (If none of the parental edges leads to an unvisited city, create an edge to a randomly chosen unvisited city.)
4. Repeat until all cities have been visited.

This operator results in approximately 30% of the edges in each parent being inherited, while the remaining 40% of the edges in the offspring do not occur in either parent.

GA E uses a variant of the PMX operator proposed by Goldberg (Goldberg 1985). Given two parents, create an offspring by the following procedure:

1. Make a copy of the second parent.

2. Choose an arbitrary subtour from the first parent.

3. Make minimal changes in the offspring necessary to achieve the chosen subtour.

For example, if the first parent is

$$(abcdefgh)$$

and the second parent is

$$(aecgbdfh)$$

and the chosen subtour is (cde), the resulting offspring is

$$(acdegbfh)$$

Notice that the offspring bears a structural relationship to both parents. The roles of the parents can then be reversed in constructing a second offspring.

Although these schemes do promote the inheritance of edges and subtours, the performance of the resulting GA's are uniformly discouraging. The trajectories for the best tour length for these two algorithms are compared with random search in Figure 4-4. While both methods outperform random search, they leave much room for improvement. The apparent cause of the poor performance is that selection pressure based on the overall tour length is not sufficient to efficiently distinguish among small competing subtours (see Theorem 1 in (Grefenstette 1985)).

Fortunately, additional knowledge concerning the TSP can be brought to bear to increase the chance of producing offspring that are improvements over their parents. One solution is to supplement selection pressure by crossover operators that use additional domain knowledge — the costs of the edges appearing in the parents — in the construction of offspring. A class of heuristic crossover operators for the TSP may be defined as follows:

1. Randomly choose a city as the current city for the offspring tour.

2. Consider the four (undirected) edges incident to the current city in the parents. Define a probability distribution over these edges based on edge cost such that the probability associated with an edge incident to a previously visited city is 0.

3. Select an edge based on this distribution. (If none of the parental edges leads to an unvisited city, create an edge to a randomly chosen unvisited city.)

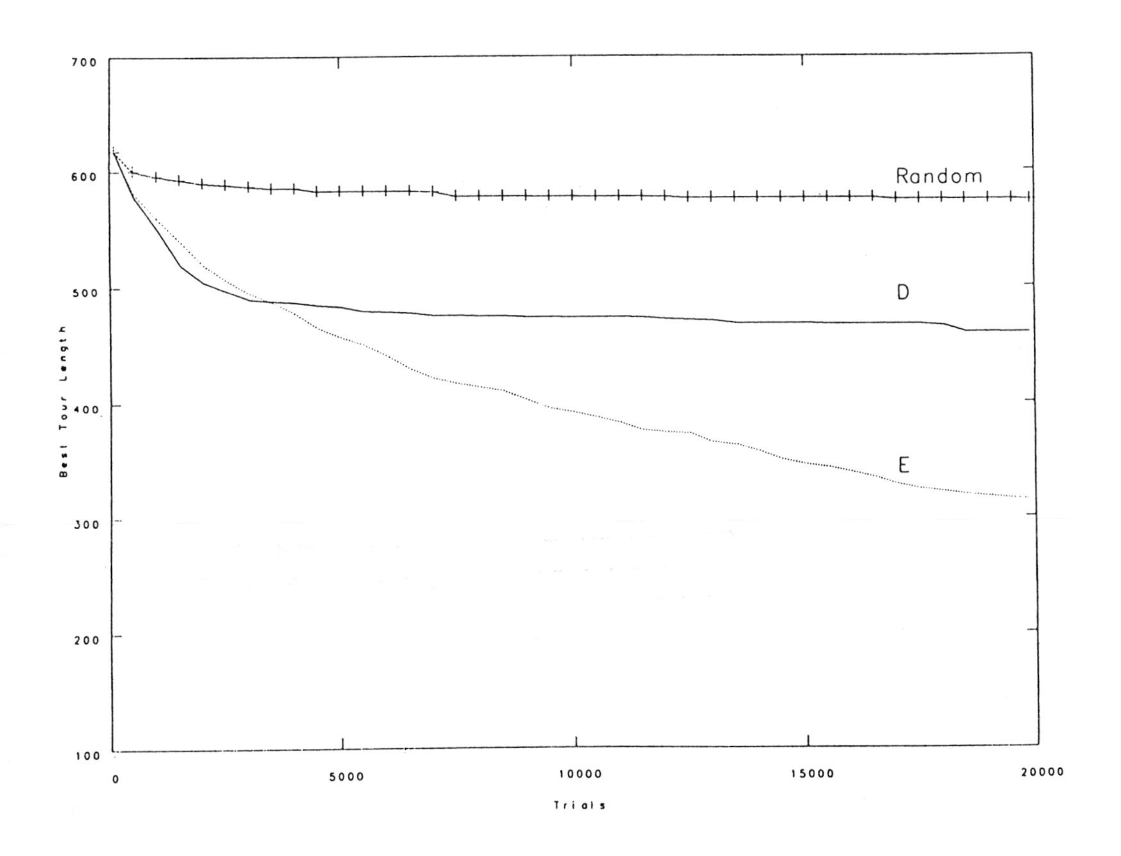

Figure 4-4: Performance of Blind Crossover Operators

Heuristic Crossover Operators			
	GA		
Problem	A	F	G
1	105.50	103.93	113.24
2	109.60	108.79	124.39
3	106.55	107.12	117.54
4	104.55	102.99	114.59
5	108.53	108.12	122.49
6	107.77	113.98	131.70
7	107.32	109.92	116.14
8	110.88	108.37	122.57
9	105.00	111.12	110.11
10	108.66	109.82	117.08

Table 4-2: Results with Heuristic Crossover

4. Repeat until all cities have been visited.

The amount of consideration given to the edge costs depends on the distribution defined in step (2). For example, GA *D* above uses a uniform distribution in step (2). This represents a base case in which no weight is given to the edge costs. We now illustrate three other possibilities for using the edge costs in forming the offspring tours. In GA *A* described previously, each parental edge is assigned a probability:

$$p_i = \frac{c_i}{\sum_{j=1}^{4} c_j}$$

where $c_i = 1/$ (the cost of edge i), for $i = 1,2,3,4$. This distribution favors short edges over longer edges, but even long edges have some prospect of being inherited. As a more extreme case, GA *F* uses a degenerate distribution which assigns probability 1 to the shortest parental edge. Yet another possibility is to let the weight given to the edge costs vary adaptively as the search progresses. As an example, we define algorithm GA *G* in which each parental edge is assigned a probability as in GA *A*, but the resulting distribution is attenuated by the population entropy so that as the entropy approaches 0 the distribution approaches a uniform distribution.

The results of using the three heuristic crossover operators are shown in table 2. GA *A* and GA *F* dominate GA *G*, which clearly loses on nine of the ten test problems. The relationship between GA *A* and GA *F* is not immediately clear. The difference in performance as indicated by the best tour length is not statistically significant, although GA *F* appears to have a slight edge. However, the performance profiles in Figures 4-5 and 4-6 are enlightening. Figure 4-5 shows the trajectories of the best tour length averaged over the 10 test problems.

As the figure shows, the deterministic crossover operator in GA F yields rapid convergence to good tours, but GA A eventually catches up. A comparison of the entropy trajectories in Figure 4-6 indicates that GA A retains the diversity required for further progress while GA F has lost any hope of further improvement.

Therefore, GA A is clearly superior to GA F when consideration is given to both efficient performance and maintenance of population diversity.

Like the results on seeding the initial population, these results support one of the philosophical axioms of the genetic approach: *Probabilistic choices are usually preferable to deterministic ones.* This principle is behind the success of genetic selection and it is not surprising to see it confirmed in the context of heuristic crossover operators.

4.5 Mutation and Local Search Heuristics

Genetic algorithms are not well suited for fine-tuning structures which are very close to optimal solutions. Instead, the strength of GA's is in quickly locating the high performance regions of vast and complex search spaces. Once those regions are located, it may be useful to apply local search heuristics to the high performance structures evolved by the GA. For example, Figure 4-7 shows the best tour found by GA A on one of the test problems.

The tour looks fairly good but has some obvious sub-optimal features. (In the Euclidean TSP the optimal tour never crosses itself.) A simple hill-climbing procedure for the TSP can be defined as follows:

1. Consider each edge in the tour;
2. Reverse the edge if this yields a shorter tour;
3. Repeat until no edges can be reversed.

The result of hill-climbing on the same tour is shown in Figure 4-8. It is clear that our simple hill-climbing procedure has arrived at a local minimum. This suggests that it may be useful to incorporate local search as a mutation operator in GA's so that the danger of getting caught in local minima is reduced by the effects of other genetic operators. GA H incorporates a mutation operator that simply reverses a randomly chosen subtour (this is called *inversion* in the GA literature (Holland 1975)). This operator serves as the basis for the 2-OPT heuristic for the TSP studied by Lin and Kernighan (Lin 1973) and is also used in the application of simulated annealing to the TSP (Kirkpatrick 1983). With a fairly low rate of mutation (occurring in 5% of the population), GA H shows a slight improvement over GA A. However, the hill climbing post processor still can be applied fruitfully to results found by GA H. Figures 4-9 and 4-10 show the results of GA H on the same problem as Figure 4-7, before and after hill-climbing, respectively. It appears that a heuristic mutation operator can achieve some of the effects of local search, but perhaps the optimal arrangement is to use a GA as a pre-processor for one of the many domain specific local search techniques for the TSP (Golden 1985).

4.6 Conclusions

The current set of experiments provide evidence for the following conclusions:

- It is important to ensure that a great deal of variety exists in the initial population to maximize the opportunities for the GA to explore the greatest possible number of feature combinations. Heuristic initialization may be helpful but must be done carefully to avoid premature convergence, since the GA is likely to exploit the opportunity to converge to the regions of the search space represented by the heuristically chosen structures.

- It is possible to define heuristic crossover operators that effectively supplement the selective pressure without sacrificing the inherent parallelism.

- Mutation operators that perform local search can reduce the need for additional postprocessing. However, it is worth considering using GA's to locate promising areas of the search space and then using efficient local methods to fine-tune the results.

There is still much additional territory to explore with respect to the application of GA's to the TSP. Additional studies are needed to measure the performance of GA's as the size of the TSP increases. Earlier studies suggest that comparable performance can be obtained with a linear increase in computational resources. Further experiments will continue to explore this possibility.

It is also worthwhile to compare the performance of GA's with other heuristic techniques suitable for very large TSP's, such as simulated annealing. This would be especially interesting since simulated annealing appears to share many characteristics of GA's. Both are probabilistic algorithms based on analogies with physical systems. Both techniques employ an adaptive criterion for accepting moves within the search space. In particular, the selective pressure in a GA appear to provide an implicit counterpart to the annealing schedule. In both cases, fluctuations in the performance of the candidate structures are are freely tolerated near the beginning of the search and less so as the system converges on high performance regions of the search space. Of course, there also interesting differences between the two techniques. One difference suggested by this study is that GA's may offer more opportunities for incorporating problem specific heuristics than simulated annealing does. This work also shows that the TSP is a natural domain for further exploration of the relationships between these two techniques.

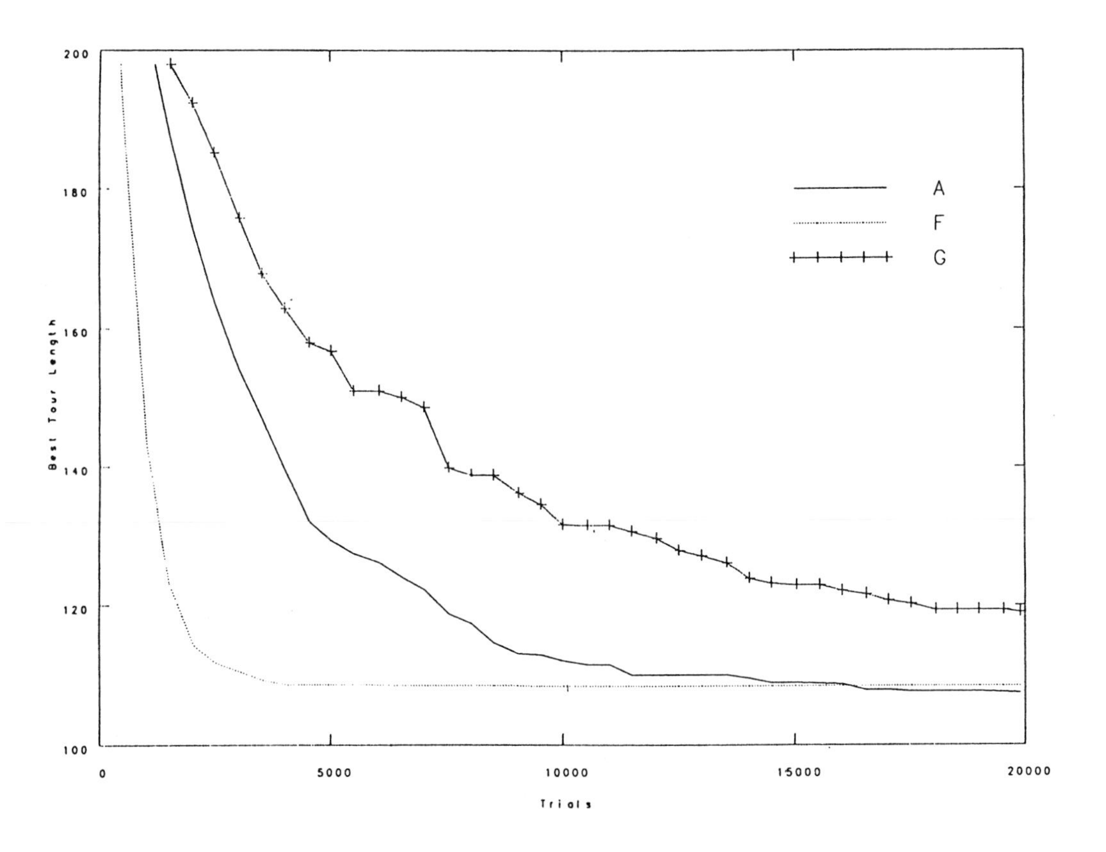

Figure 4-5: Performance of Heuristic Crossover Operators

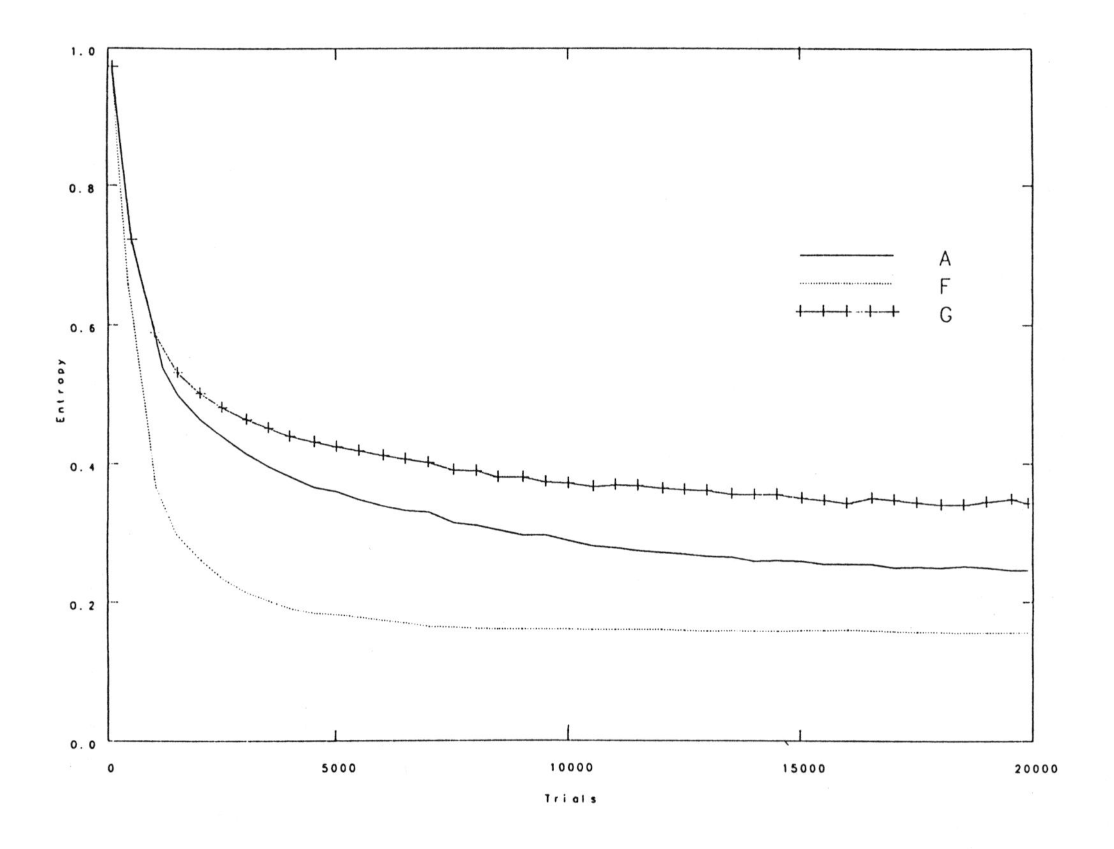

Figure 4-6: Convergence of Initialization Methods

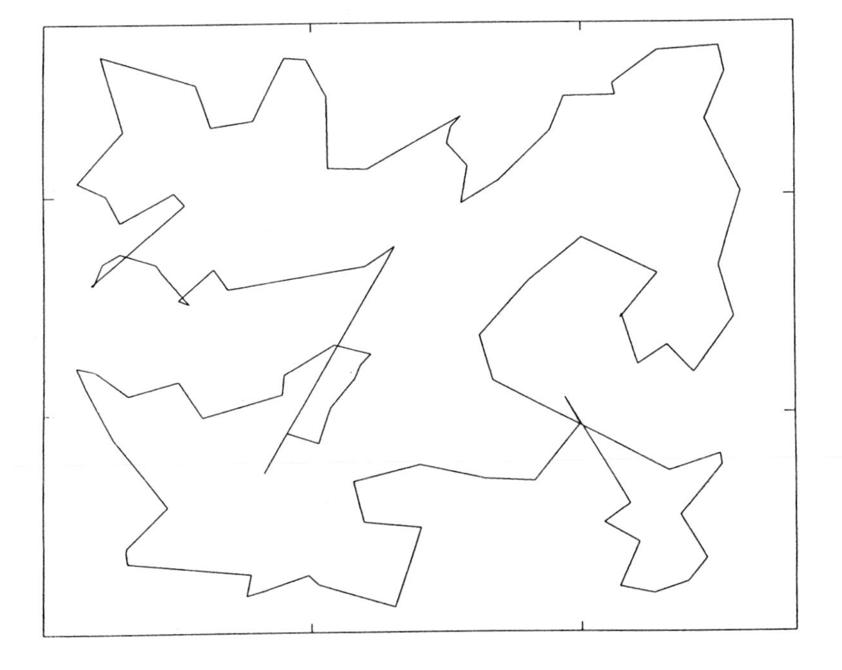

Figure 4-7: Typical Best Tour, No Mutation

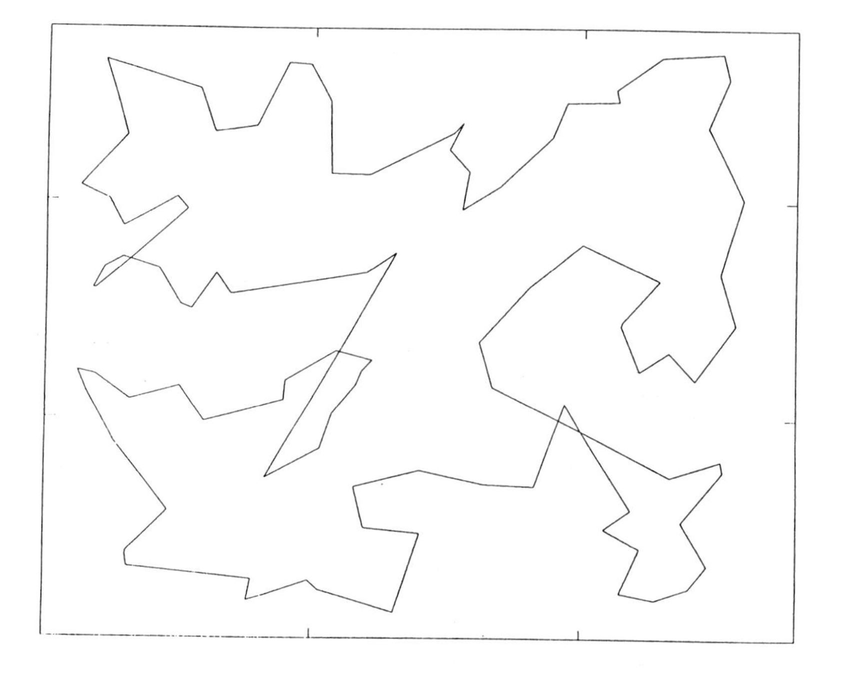

Figure 4-8: Best Tour After Hill-Climbing

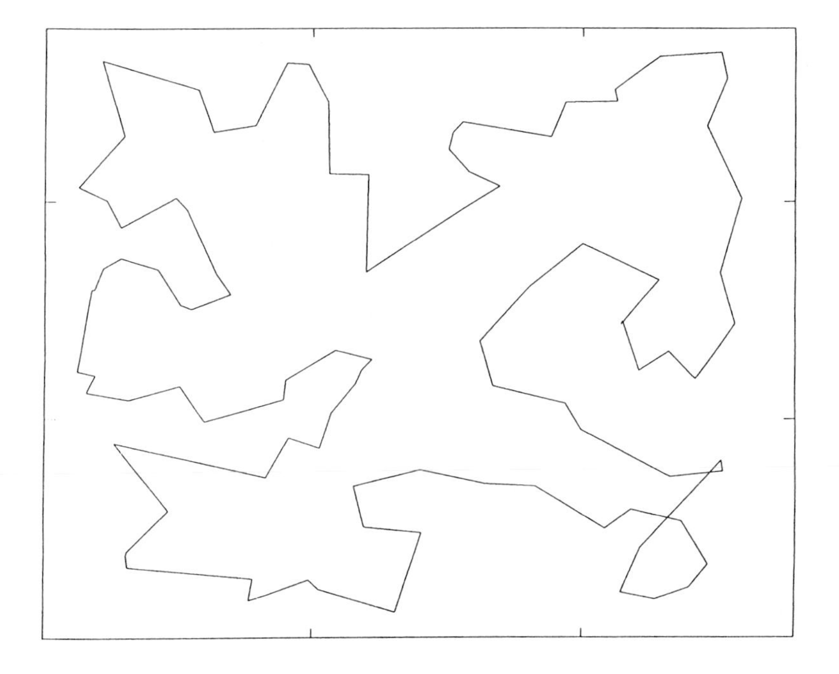

Figure 4-9: Typical Best Tour with Mutation

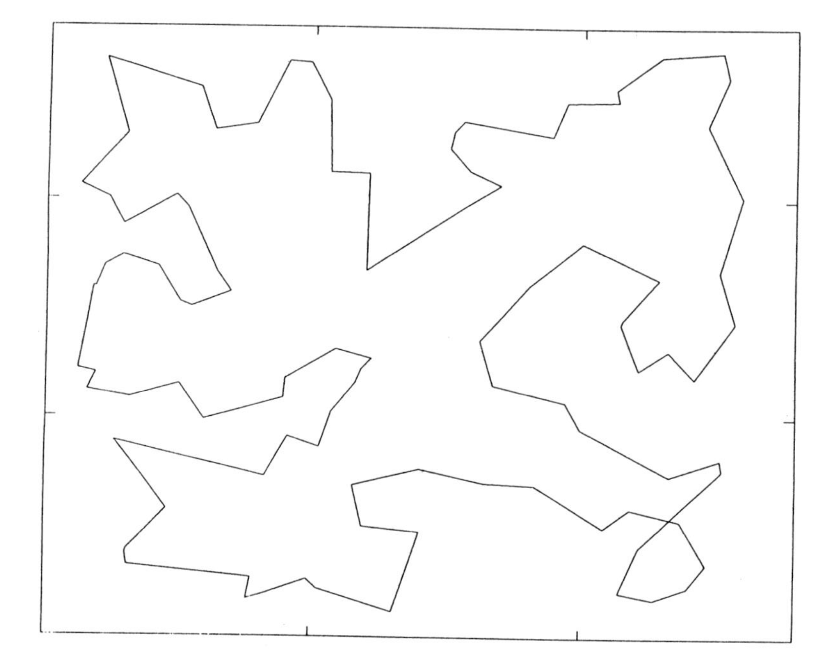

Figure 4-10: Same Tour After Hill-Climbing

Lashon Booker

Chapter 5
Improving Search in Genetic Algorithms

Abstract

Genetic algorithms have been touted as a class of general purpose search strategies that maneuver through complex spaces in a near optimal way. Implementations of these algorithms, however, have been prone to converge prematurely before the best solution has been found. This paper shows that carefully chosen modifications to the way search operators are implemented can be effective in alleviating this problem.

5.1 Introduction

Searching a complex space of problem solutions often involves a tradeoff between two apparently conflicting objectives: exploiting the best solutions currently available and robustly exploring the space. Making and testing hypotheses about how to improve a good solution "exploits" that solution in the sense that improved variants are more likely to be found. This is the most direct way to take advantage of the available knowledge. On the other hand, examining regions of the search space not yet explored might provide knowledge that leads to the discovery of even better solutions. An emphasis on robust exploration thereby reduces the likelihood that promising regions in the space will be overlooked. Hill climbing is a good example of a search strategy that exploits the best among known possibilities for finding an improved solution. Because hill climbing strategies take such a limited view of where the opportunities for improvement are, they are vulnerable to getting trapped in regions of the space that are far removed from the optimal solution. Random search is a good example of a search strategy that is concerned only with robust exploration. Because this strategy ignores what is known about the space and blindly samples every region, it is usually intolerably inefficient.

Genetic algorithms have been touted as a class of general purpose search strategies that strike a reasonable balance between exploration and exploitation. Indeed, the theoretical analysis of genetic algorithms suggest that they manage this tradeoff in a near optimal way (Holland 1975). The power of these algorithms is derived from a very simple heuristic assumption: that the best solutions will be found in regions of the search space containing relatively high proportions of good solutions; and, that these regions can be identified by judicious and robust sampling of the space. Holland shows how simple mathematical models of population genetics can efficiently and implicitly make use of this heuristic. Genetic algorithms implement these models by iteratively manipulating a population of strings using selection together with search operators like crossover and mutation. The algorithms have been successfully applied to several kinds of problems, including function optimization (DeJong 1975, Smith and DeJong 1981), adaptive control (Goldberg 1985), game playing (Mauldin 1984, Smith 1983), cognitive modeling (Booker 1982, Wilson 1986a), and image understanding (Fitzpatrick, Grefenstette, and Van Gucht 1984).

Despite the demonstrated advantages of these algorithms, the feeling persists that there is much to learn about how to effectively implement a genetic algorithm. While the performance of most implementations is comparable to or better than the performance of many other search techniques, it still fails to live up to the high expectations engendered by the theory. The problem is that, while the theory points to sampling rates and search behavior in the limit, any implementation uses a finite population or set of sample points. Estimates based on finite samples inevitably have a sampling error associated with them. Repeated iterations of the algorithm compound the sampling error and lead to search trajectories much different from those theoretically predicted. This problem is manifested in practice as a premature loss of diversity in the population with the search converging to a sub-optimal solution.

Many remedies for combatting these finite stochastic effects have been proposed. The importance of this problem was first noted in DeJong 1975. DeJong showed that bounding the error in the selection probabilities for each string substantially improves the performance of the algorithm. Others have tried reducing error by using more accurate selection procedures (Brindle 1981) and by optimizing the algorithm's parameters (Grefenstette 1986), all with mixed results. While intuitively these changes should be beneficial, they have not resulted in a version of the algorithm that is reliably better in general. A recent trend in response to this problem has been to violate the tenets of the theory whenever an improvement in performance can be achieved (e.g. Baker 1985, Mauldin 1984). This paper takes the position that the desired improvements in performance are available within the framework of the theory. By carefully implementing the search operators in genetic algorithms, the balance between exploration and exploitation can be adjusted to help avoid premature convergence.

5.2 Overview Of The Theory

In order to show that control of search is an important issue, we begin with a brief overview of the relevant theory. This provides the basis for an analysis of the causes of premature convergence.

Genetic algorithms are derived from a simple model of population genetics based on the following assumptions:

1. Chromosomes or genotypes are fixed length strings having a finite number of pos-

sible values or *alleles* at each position.

2. A population contains a finite number of genotypes.

3. Each genotype has a fitness, or relative ability to survive and produce offspring.

The population of strings is a dynamic entity. During each iteration of the algorithm — a *generation* — the fitness of every string is computed, and strings are stochastically chosen to become parents and reproduce according to their relative fitness. Variation is introduced into the new strings using crossover and mutation, after which the new strings replace existing ones and the next generation begins. The sequence of populations generated by the algorithm constitutes a parallel search trajectory through the space of all possible strings. The genotypes that survive will, over time, be those which have proven to be the most fit. In this sense, the search trajectory is steered toward those types of individuals, or regions of the search space, with above average fitness.

The driving force behind a genetic algorithm is the reproduction of individuals in proportion to fitness together with the crossover operator[1]. Reproduction according to fitness concentrates the search in regions of high observed average fitness. Crossover is a search operator that probes both familiar and unexplored regions of the search space with each application. To make this explanation more precise, consider a particular set of regions in the search space called hyperplanes or *schemata* (Holland 1975). A schema is the set of all strings having certain "defining" values at designated positions in the string. For example, the set of all strings having the value 0 in the first position is a schema having only one defining value. This schema can be compactly specified by the string $0***...***$, where the asterisks indicate positions not important to the definition of the schema. Strings belonging to the region of the search space designated by a schema are said to be *instances* of that schema. Every string of length k is an instance of exactly 2^k distinct schemata.

By formally treating each schema as a random variable whose mean is estimated by the average fitness of its instances in the population, genetic algorithms can be mathematically analyzed (Bethke 1981, Holland 1975). Three basic properties of genetic algorithms have emerged from this analysis:

1. In a population of size M, the estimated values of approximately $M^2 2^{k/2}$ schemata are adjusted every generation. This information is used by the algorithm to increase or decrease the number of instances of every schema according to whether it is above or below average. More precisely, if $\mu_s(t)$ is the average value of the instances of schema s at generation t and $n_s(t)$ is the number of those instances in the population, then

[1]Mutation is only a background operator that guarantees no allele will be permanently lost from the population.

$$n_s(t+1) = \left[\frac{\mu_s(t)}{\hat{\mu}}\right](1-\epsilon)n_s(t)$$

where $\hat{\mu}$ is the average fitness of all strings in the population and $\epsilon < 1$ is a very small positive number. This evaluation and management of a large number of schemata is accomplished by simple manipulations of only M strings at a time, a phenomenon Holland calls *intrinsic parallelism*.

2. Iterative application of the genetic algorithm leads to a population in which the number of instances of each schema is proportional to its observed average value. In this sense, the population is a database that compactly and usefully summarizes the information obtained from the search. The genetic algorithm automatically generates and utilizes this database.

3. The set of all schemata is a rich and highly redundant covering of the entire search space. It can be proven (Bethke 1981) that the algorithm will consistently discover and exploit highly rated schemata for a large class of fitness functions. From the standpoint of the theory, genetic algorithms are not easily misled.

5.3 The Premature Convergence Problem

It is clear from the theory that as the proportions of better schema increase in the population, the proportions of less desirable schema will decrease. Eventually the very best schemata and individuals occur in dominating proportions and, in this sense, the search converges to a solution. What remains to be determined is why this convergence often happens *prematurely* in implementations of the algorithm, before the optimal solution is found.

It is impossible for a parent string to produce anything but an integral number of offspring. It is also not possible for the number of instances of a schema to reflect the desired proportion with arbitrary precision in a finite population. This is an inherent source of sampling error which, coupled with the variance associated with stochastic selection processes, can lead to wide deviations in the number of instances of a schema from what is theoretically predicted. The effect of such errors can be cumulative to the point where useful schemata and their alleles may disappear from the population. This phenomenon is called *genetic drift* and is a well-studied problem in population genetics. It has long been suspect as a primary cause of premature convergence in genetic algorithms (DeJong 1975). Genetic drift can be reduced by taking measures like increasing the population size. Unfortunately, this can only be done in moderation since it eventually makes the search too inefficient. It is therefore important to investigate and eliminate other causes of premature convergence.

5.3.1 Loss of Diversity

At first glance it would seem that premature convergence is best treated by an increase in the mutation rate to restore the lost alleles. As DeJong pointed out, however, higher mutation rates will disrupt the proliferation of high performance alleles as well as the poor ones. Mauldin (Mauldin 1984) proposes a clever variation on this theme of enforcing diversity to avoid premature convergence. He begins by defining a *uniqueness* value as the minimum Hamming distance allowed between any offspring and all existing strings in the population. Whenever a new individual is closer than this to an existing structure, the allele values which match are randomly changed in the offspring until the required distance is achieved. This amounts to a kind of knowledge-directed increase in the mutation rate. To make sure that the genetic algorithm is allowed to eventually converge, Mauldin slowly decreases the uniqueness value from k to 1 as the search proceeds, using the relation

$$\textit{Hamming distance} > k_t = \left\lceil \frac{k(N-t)}{N} \right\rceil$$

where N is the total number of generations and k_t is the uniqueness value at generation t. Mauldin shows that uniqueness improves the rate at which the search converges toward a good solution.

There are two problems with the uniqueness approach. First, it can be prohibitively expensive to implement if the population is large or the strings are long. A more important concern, though, is that heavy-handedly forcing the genetic algorithm to sample the space robustly prevents parent strings from reliably transmitting good schemata to their offspring. The statistical summary of the value of each schema that the algorithm maintains is thereby destroyed. Directly manipulating the diversity in a population alleviates the primary symptom of premature convergence without providing any insights about the causes or how to eliminate them.

5.3.2 Predicting Rapid Convergence

One strategy for finding the causes of premature convergence in genetic algorithm implementations is to look for indicators that occur before loss of diversity. If the indicators have a non-spurious correlation with the loss of diversity, they are likely to reveal a causal relationship. Baker (Baker 1985) observed that rapid convergence often occurs after an individual or small group of individuals contributes a large number of offspring to the next generation. Since populations are finite, a large number of offspring for one individual means fewer offspring for the rest of the population. When too many individuals get no offspring at all, the result is a rapid loss of diversity and premature convergence. By measuring the percentage of the current population producing offspring — a measure Baker calls the *percent involvement* — one can anticipate this rapid convergence and have a chance to prevent it.

The preventive strategies Baker investigates are concerned with selection methods and how the number of offspring assigned to each string is computed. The most effective method he discusses is to deterministically compute the expected number of offspring for each string based on the rank of its fitness in the population, not the magnitude of the fitness. Given a target value for the number of offspring for the best string, Baker assigns linearly decreasing values to the rest of the population so that the total number of offspring is as desired. Using rank in this way allows the algorithm to maintain a relatively steady percent involvement. As a result, loss of diversity is negligible. The effects on performance are mixed. There is some increased ability to discover good solutions simply because the search is never hampered by lack of alleles and continues longer. However, the rate at which better solutions are generated is relatively slow.

While the ranking method prevents diversity from being lost too quickly, it also does not allow allele proportions to change when warranted. The algorithm's ability to exploit promising schemata is compromised by ignoring all knowledge about the relative fitness of strings in the population. Making the expected number of offspring proportional to relative fitness is at the heart of what makes genetic algorithms work in theory. To ignore this is to abandon any claim that the search is intrinsically parallel.

5.3.3 Inadequate Search

Percent involvement is a good predictor of rapid convergence. The ranking method forces percent involvement into a pre-determined range. What is needed to handle premature convergence within the context of the theory is a way to *prevent* a low percent involvement. Once again, there is a need to step back and look for causes.

The key seems to be the role relative fitness plays in the balance between exploration and exploitation. A given individual will have a high expected number of offspring only if it has a high fitness relative to the rest of the population. This fitness advantage is what causes the drop in percent involvement and the subsequent loss of diversity. The loss of diversity is premature only when the fitness advantage is misleading; that is, when the context of the current population makes a string look better than it really is. The fundamental issue, therefore, is the extent to which the population is a representative sample of the best regions in the search space.

The only way to assure that a sample is representative is to make sure that the exploration of the space is as thorough as possible. In genetic algorithms this exploration is done by the crossover operator. It is therefore reasonable to hypothesize that improvements in the way crossover is implemented will help alleviate the premature convergence associated with low percent involvement. The remainder of this paper tests this hypothesis by proposing and testing several alternative ways of implementing crossover.

TABLE 1 - Standard Set of Test Functions

Function	Search Space: Bounds	Search Space: Resolution	Description
$f1 = \sum_{i=1}^{3} x_i^2$	$-5.12 \le x_i \le 5.12$	$\Delta x_i = 0.01$	3-D parabola
$f2 = 100\,(x_1^2 - x_2)^2 + (1-x_1)^2$	$-2.048 \le x_i \le 2.048$	$\Delta x_i = 0.001$	Non-convex deep parabolic valley
$f3 = \sum_{i=1}^{5} \lfloor x_i \rfloor$	$-5.12 \le x_i \le 5.12$	$\Delta x_i = 0.01$	5-D step function
$f4 = \sum_{i=1}^{30} i x_i^4 + GAUSS(0,1)$	$-1.28 \le x_i \le 1.28$	$\Delta x_i = 0.01$	Quartic with Gaussian noise
$f5 = \left[\frac{1}{500} + \sum_{j=1}^{25} \frac{1}{j + \sum_{i=1}^{2} (x_i - a_{ij})^6} \right]^{-1}$	$-65.536 \le x_i \le 65.536$	$\Delta x_i = 0.001$	Plateau with 25 deep perforations centered at (a_{1j}, a_{2j})

5.4 Making Search More Efficient

Three modifications to the search operator will be suggested. Following the standard procedure for evaluating genetic algorithms (DeJong 1975, Grefenstette 1986), each modification will be tested on a function minimization task using the five functions described in Table 1. The functions are all normalized to the same range [0,100] to help guarantee consistent comparisons. Performance will be evaluated over an interval of 6000 function evaluations and analyzed using the traditional measures: *on-line, off-line,* and *best-so-far.* On-line performance is simply the mean of all function evaluations and measures the average quality of the points sampled. This measure is given by

$$On\text{-}line\ (T) = \left(\frac{1}{T}\right) \sum_{i=1}^{i=T} f_E(i)$$

where $f_E(i)$ is the average of the ith function value for all five functions and T is the total number of evaluations. In accordance with Bethke's suggestion (Bethke 1981), the components of f_E are weighted by the performance of random search on each function. Off-line performance is a more traditional measure for a function optimizer, monitoring the convergence of the search toward the optimum. It is given by

$$Off\text{-}line\ (T) = \left(\frac{1}{T}\right) \sum_{i=1}^{i=T} f_E^*(i)$$

where $f^*_E(i)$ is the weighted average of the best value found for each function after i evaluations. The best-so-far measure is simply f^*_E.

5.4.1 Two Crossover Points

The first modification to crossover focuses on the number of crossover points involved. Crossover is usually implemented by choosing one crossover point at random, then exchanging segments between the two parent strings. Each application of the crossover operator searches both familiar and unexplored regions of the search space. This can easily be seen by examining the way crossover manipulates schemata. Consider the two binary strings 100101010 and 010101101. If crossing over occurs between the third and fourth positions, the resultant strings will be 100101101 and 010101010. These new strings are both instances of schemata like $**0101***$ that were represented by the parent strings. The genetic algorithm uses these new instances to revise its estimates of the average fitness of such schemata. The new strings are also instances of schemata like $10*****01$ and $01*****10$ which were not represented by either parent string. The genetic algorithm uses these new schemata to broaden its search and increase the likelihood that all important regions of the search space are sampled. One side effect of this recombination is that some schemata present in the parent strings are "disrupted" and not transmitted to the offspring. In the current example the schemata $10*****10$ and $01*****01$ are both lost.

The probability that a schema ξ will be disrupted by crossover is simply $l_\xi/(l-1)$ where l_ξ is the distance between the outermost defining alleles of ξ and l is the length of the string[2]. This disruption does not in theory seriously impair the search capability of genetic algorithms. In practice, though, any reduction in this disruptive effect should improve the efficiency of crossover. DeJong (DeJong 1975) noted that such a reduction should be attainable by a simple modification to the crossover operator:

> If we think of the chromosome as a circle with the first gene immediately following the last then it becomes immediately clear that there are in fact 2 crossover points: one fixed at position zero and the other randomly selected. An immediate generalization to the present crossover operator is to allow both crossover points to be randomly selected. (p. 149)

The advantage to be gained is that schemata having large l_ξ values will have a smaller chance of being disrupted by crossover.

[2] l_ξ is called the *"definition length"* of ξ.

DeJong's experiments with this idea did not show the expected improvements in performance. Subsequent investigations by Booker (Booker 1982) discovered and corrected a flaw in DeJong's implementation and demonstrated that two crossover points can indeed improve performance. This issue is revisited here because the result has never been rigorously confirmed. Accordingly, an experiment was designed to thoroughly test the advantage of using two crossover points. The algorithm was run 10 times on each test function to obtain an average performance estimate. This procedure was repeated at least 10 times to compute statistics about the expected overall performance. Statistical significance is determined by using a t test with $\alpha = 0.5$ comparing the means of the two groups. This experimental procedure is used for all results presented in this paper.

The results of this experiment were conclusive. On-line performance was significantly poorer ($t = 2.245$, $df = 23$) using two crossover points[3]. This is to be expected however since on-line performance is best served by favoring exploitation over exploration and using two crossover points enhances exploration. Off-line performance, on the other hand, is dramatically improved ($t = 4.322$) as is best-so-far ($t = 2.575$). The importance of this result is that the genetic algorithm now shows an improved ability to sustain a robust search after the initial discovery of promising schemata. This can be seen more clearly by modifying the off-line measure with increasing weights $w_{t+1} > w_t$ so that it emphasizes convergence over initial performance. Using exponentially increasing weights chosen to make the effects of the initial population negligible after 6000 trials[4], we see a significantly improved ability to continue progressing toward the optimum ($t = 3.315$). It should be noted that this one change to the crossover operator already gives improvements that match or exceed those achieved by Mauldin with his uniqueness method (The un-normalized, unweighted values are -2.606 off-line and -5.34 best-so-far).

5.4.2 Improved Crossover Implementation

Clearly, the ability of a genetic algorithm to maintain an effective search depends on the continued ability of the crossover operator to perform the search functions previously described. Unfortunately, what happens in practice is that crossover becomes less effective over time as the strings in the population become more similar. One factor that might contribute to premature loss of diversity is a needlessly ineffective crossover operation. When two strings are different but the crossover points fall so as to exchange an identical segment, the resultant strings will be identical to the original ones. This is a lost opportunity to sample new schemata. Theoretically we can expect these opportunities to occur again. In practice, though, the premature loss of diversity may preclude future sampling of these regions.

[3]The notation in parentheses is used to show the value of the t statistic and the number of degrees of freedom involved.

[4]The weighted off-line measure is given by *Off-line* $(T) = (\frac{1}{T})\sum_{i=1}^{i=T} w^{T-i} f^*_E(i)$, where $w = 0.9997$.

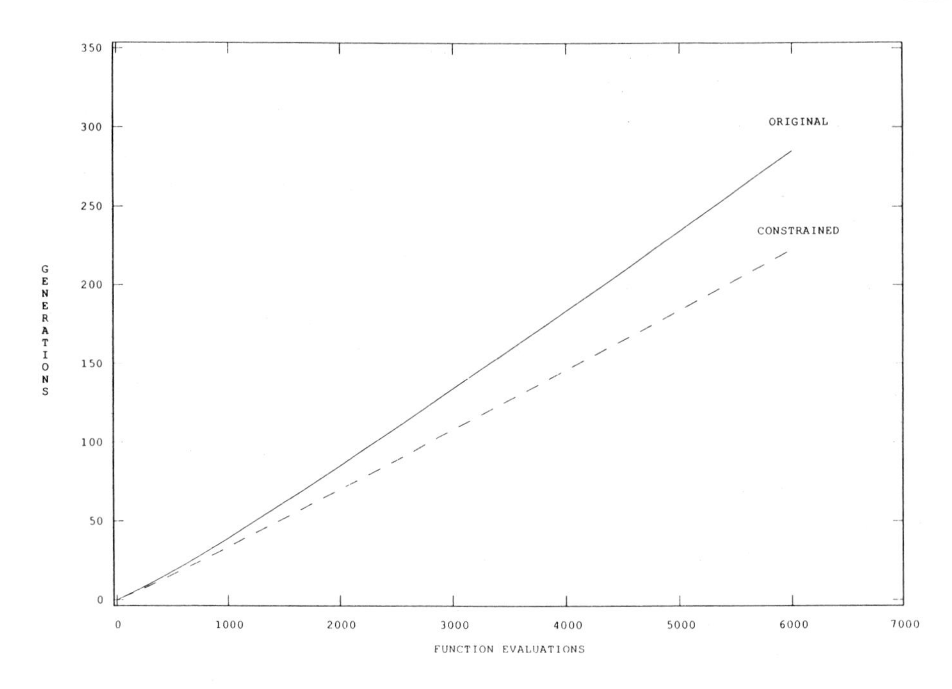

Figure 5-1: Efficiency of the Improved Crossover Operator on Test Function f1

Consequently, we speculate that the performance of a genetic algorithm can be improved if crossover is constrained to always produce variations whenever possible. This constraint can be implemented by simply restricting the location of crossover points. For example, given the two strings 010110100 and 001000100, at least one crossover point should be chosen between positions one and four. More generally, given any two strings we examine their *"reduced surrogates"* containing only the non-matching alleles. The reduced surrogates of the previous strings are 1011 and 0100. Crossover points are randomly selected for these reduced strings, then mapped back into the original strings.

Experiments with this new crossover operator show that it significantly degrades on-line performance ($t = 6.566$, $df = 18$), but this is to be expected as noted previously. Best-so-far performance is more or less unaffected. Off-line performance is clearly improved but not significantly so ($t = 1.272$). This is surprising since the new operator is decidedly more efficient in that fewer generations are required to do a given number of function evaluations (see Figure 5-1). One explanation for the less than spectacular improvement is that there is insufficient "grist for the mill" in the early generations. Any implementation of the genetic algorithm must be careful that the initial population contains an adequate pool of alleles for each position. It might be equally important that there be an adequate initial pool of schemata! If the initial distribution of points does not uniformly sample the space, crossover may never get the opportunity to direct the search into the unrepresented regions. The commonly used multiplicative random number generators are very poor when it comes to distributing samples randomly in more than one dimension (Marsaglia 1968). Random number generators having this capability are computationally expensive. A good compromise is to produce initial populations as usual, then subject them to repeated crossing over with uniform random pairing. If this is done to the point of stochastic equilibrium[5] we can be assured that the initial pool of schemata will be as robust a sample as possible. This change has the anticipated effect. It improves the off-line performance of the new crossover operator to the point where it is significantly better than the previous one ($t = 2.098, \quad df = 21$).

5.4.3 Variable Crossover Rate

The final aspect of crossover to be considered is the frequency with which it is applied. Most implementations keep the crossover rate fixed throughout the interval of observation. Recent work by Wilson (Wilson 1986a) on genetic algorithms for classifier systems[6] suggests that dynamically varying the crossover rate may be beneficial. Wilson defines an entropy measure over the population and adjusts the crossover rate up slightly if entropy is falling and down if entropy is flat or rising. The idea is to go easy with crossover until variation has been "absorbed", then introduce more variation by increasing crossover again. This method seems to improve the performance of genetic algorithms in a classifier system. However, it is questionable how appropriate it is for more standard genetic algorithms. Entropy basically indicates diversity and we have already seen that by the time diversity has decreased it may be too late to halt premature convergence. The sudden onset of allele loss is not a problem in classifier systems because the fitness of a string is averaged over several trials and the genetic algorithm is used sparingly.

Nevertheless, the idea of varying the crossover rate is appealing. Rapid loss of diversity can be viewed as a sudden change in which exploitation has the upper hand over exploration. The harmful effects of this phenomenon have already been discussed. It is also detrimental though if exploration proceeds faster than exploitation can assimilate the new information. When the fitness values in the population are close together, for instance, it will take time for the selective pressures to sort out the best from the worst and adjust the proportions of schemata accordingly. Introducing a lot more variation at this point is counterproductive. A variable crossover rate is one way to compensate for these shifts in the balance between exploration and exploitation. The idea is similar to

[5]This means that the probability of occurrence for any schema is well-defined and given by the fixed point of the stochastic transformation. In the steady state distribution the probability of occurrence of every schema is the product of the proportions of its defining alleles. Holland (Holland 1975) shows that even severe departures from equilibrium will be substantially reduced after each string has been crossed at almost every position.

[6]A classifier system is a special kind of production system designed to permit non-trivial modification and reorganization of its rules. In particular, the genetic algorithm can be used as a learning heuristic. See Holland 1986 for more details.

TABLE 2 - Performance Summary
(Numbers in parentheses indicate that the difference in performance between this implementation and the previous one is **not** statistically significant)

Crossover Implementation	Normalized Performance Values			
	On-Line	Off-line	Best-so-far	Weighted Off-line
One crossover point	3.3694	1.7260	0.1444	0.9014
Two crossover points	3.5340	1.4119	0.0265	0.6739
Constrained crossover	4.6410	1.2884	(0.0304)	(0.5888)
Variable crossover rate	4.1372	1.1540	(0.0238)	0.5137

the way search is controlled in simulated annealing except that it is automatic instead of following a fixed schedule.

A good measure for predicting allele loss is percent involvement. It is possible that by changing the crossover rate in response to changes in percent involvement, performance can be improved. Accordingly, a simple variation on this theme is proposed. Every percentage change in percent involvement will be countered with an equal and opposite percentage change in the crossover rate. To guarantee that search never becomes completely ineffective, the crossover probability is not allowed to drop below 0.2. Experiments with this modification to crossover confirm the hypothesis. On-line performance ($t = 3.093$, $df = 21$) and off-line performance ($t = 2.813$) are both significantly improved. The algorithm also shows an improved ability to sustain its search. Looking again at the weighted off-line measure we see that varying the crossover rate provides another significant improvement in this capability beyond the first one ($t = 3.018$, $df = 18$).

5.5 Conclusion

Table 2 summarizes the outcome of all the experiments described here. These results show that carefully chosen modifications to the way search operators are implemented can dramatically improve the performance of genetic algorithms. The modifications were all straightforward: using two crossover points instead of one; implementing crossover more carefully so that it introduces variation whenever possible; and varying the crossover rate dynamically to compensate for imbalances in the exploration versus exploitation tradeoff.

The performance improvements realized from these changes exceed those achieved by others using changes to the algorithm that violate the tenets of the underlying theory.

It is hoped that these results convincingly demonstrate that the theory of genetic algorithms is the proper point of departure for devising improved implementations of genetic algorithms.

David E. Goldberg

Chapter 6
Simple Genetic Algorithms and the Minimal, Deceptive Problem

6.1 Introduction

Ever since Holland unveiled the power of genetic algorithms (GA's) through the theory of schemata (Holland 1968, Holland 1975), researchers have understood that genetic algorithms work because above-average, short, low-order schemata (these have been called building blocks) receive exponentially increasing numbers of trials in successive generations. This near-optimal sampling rate (in the sense of a multi-armed bandit problem) is especially propitious when we realize that the building blocks are mixed and matched through the action of structured, yet randomized recombination operators (crossover), and near-optimal alternatives are quickly located. Many empirical studies of genetic algorithms in search and machine learning applications (Goldberg and Thomas 1986) have demonstrated this desirable result; however, few studies have gone beyond Holland's original efforts and examined the detailed, theoretical underpinnings of the simple GA.

In this paper, we examine the expected performance of a simple genetic algorithm in a simple problem especially designed to give the GA difficulties. This problem, which we call the minimal, deceptive problem (MDP), intentionally tries to mislead the simple genetic algorithm away from the optimal point and toward sub-optimal solutions. It is a surprising result that despite our best efforts, the GA usually converges to the correct answer (the global optimum). In other words, even though the problem is intentionally misleading, the simple genetic algorithm usually refuses to take the bait and converges on the global optimum instead.

In the remainder, we review existing theory, motivate the minimal, deceptive problem, perform an extended schema analysis, and consider the convergence or divergence of a simple genetic algorithm with reproduction and crossover only. We also consider extensions of these results to higher order problems and problems where reordering operators may be necessary.

6.2 Brief Review Of Existing Theory

The cornerstone of all genetic algorithm theory is the realization that GA's process schemata (schema-singular, schemata-plural) or similarity templates. Suppose we have a finite binary string of length 1, and suppose we wish to describe a particular similarity. For example, consider the two strings A and B as follows:

$$A = 10111$$

$$B = 11100.$$

We notice that the two strings both have 1's in the first and third position. A natural shorthand to describe such similarities introduces a wild card or don't care symbol, the star $*$, in all positions where we are disinterested in the particular bit value. For example, the similarity in the first position can be described as follows:

$$1 * * * *.$$

Likewise, the similarity in the third position may be described with the shorthand notation

$$* * 1 * *$$

and the combined similarity may be described with $*$'s in all positions but the first and third:

$$1 * 1 * *.$$

Of course, these schemata or similarity templates do not only name the strings A and B. The schema $1 * * * *$ describes a subset containing $2^4 = 16$ strings, each with a 1 in the first position. The more specific schema $1 * 1 * *$ describes a subset of $2^3 = 8$ strings, each with a 1 in both the first and third position.

More generally, we notice that not all schemata are created equal. Some are more specific than others. We call the specificity of a schema H (its number of fixed positions) its order, $o(H)$. For example, $o(1 * * * *) = 1$ and $o(1 * 1 * *) = 2$. Some schemata have defining positions spaced farther apart than others. We call the distance between a schema's outermost defining positions its defining length, $\delta(H)$. For example, the defining length of any one-bit schema is 0: therefore

$$\delta(1 * * * *) = \delta(* * 1 * *) = 0.$$

On the other hand, the defining length of our order-two schema example may be calculated by subtracting the position indices of the outermost defining positions:

$$\delta(1 * 1 * *) = 3 - 1 = 2.$$

These properties are useful in the fundamental theorem of genetic algorithms, otherwise known as the schema theorem. Under fitness proportionate reproduction, simple crossover, and mutation, the expected number of copies m of a schema H is bounded by the following expression:

$$m(H, t+1) \geq m(H, t) \, \frac{f(H)}{\bar{f}} \left[1 - p_c \, \frac{\delta(H)}{l - 1} - p_m \, o(H) \right].$$

In this expression, the factors p_m and p_c are the mutation and crossover probabilities respectively, $\overline{f}$ is the average fitness of the population, and the factor $f(H)$ is the schema average fitness which may be calculated by the following expression:

$$f(H) = \frac{\sum_{s_i \in H} f(s_i)}{m(H,t)}.$$

The schema average fitness $f(H)$ is simply the average of the fitness values of all strings s which currently represent the schema H. Overall, the schema theorem says that above average, short, low-order schemata are given exponentially increasing numbers of trials in successive generations. Holland in (Holland 1975) has shown that this is a near-optimal strategy when the allocation process is viewed as a set of parallel, overlapping, multi-armed bandit problems. We will not review this matter in detail here. Instead, we need to look at what the schema theorem doesn't tell us.

The bounding theory is useful up to a point. After all, it does assure us that we are allocating our trials (our experimental function evaluations) in an intelligent way. Furthermore, the schema theorem tells us that this intelligent processing of schemata proceeds in a highly parallel fashion. Holland has estimated that on the order of n^3 schemata (where n is the population size) are usefully processed each generation even though we only manipulate on the order of n strings. This leveraged processing of schemata is so important that Holland has given it a special name, *implicit parallelism* . Yet, despite the intelligent allocation of trials, and despite the highly leveraged processing, something seems to be missing. In fact, there is a hidden article of faith buried within the schema theorem. In an arbitrary problem with an arbitrary coding, how do we know that a simple genetic algorithm will combine short, low-order, above-average schemata (building blocks) to form near-optimal strings? I have called the assumption that this does occur a *building block hypothesis* . In general, however, we don't know beforehand that building blocks will indeed lead to the best points. It is no trivial consolation that GA's have performed well in a variety of applications, across a broad spectrum of disciplines. Additionally, it is no small bonus that we are following natural example; our own human existence is evidence that these (or similar) procedures search complex spaces quickly, blindly, and successfully. Nonetheless, we must acknowledge that the schema theorem by itself does not guarantee convergence in arbitrary problems.

Actually, the state of our theoretical knowledge about genetic algorithms is not as limited as this. Bethke's recent study in (Bethke 1981) has examined some sufficient conditions for simple GA convergence using Walsh function analysis of schema averages. This work is particularly important because it has quantified the deception required to fool a simple GA. Specifically, Bethke has shown that for a problem to have a chance of beating a simple GA, it must have misleading building blocks. Good, low-order, short schemata must mix and match to find bad strings. Although Bethke was able to give examples of GA-hard functions, he was unable to characterize them generally.

Furthermore, Bethke's Walsh-schema analysis is not particularly useful in practical GA search because a complete analysis is as at least as burdensome as an enumerative search of the underlying space. Additionally, the Walsh-schema analysis, although analytically useful, removes us from the essence of the deception required to fool a simple genetic algorithm. This is why we introduce the minimal, deceptive problem (MDP) in the next section.

6.3 The Minimal, Deceptive Problem

Let's construct the simplest problem that should cause a GA to diverge from the global optimum. To do this, we want to violate the building block hypothesis in the extreme. Put another way, we would like to have short, low-order building blocks lead to incorrect (sub-optimal) longer, higher-order building blocks. The smallest problem where we can have such deception is in a two-bit problem. Suppose we have a set of four, order-two schemata over the same two defining positions, each schema associated with a fitness value as follows:

$$\begin{array}{cccccc} *** & 0 & ***** & 0 & * & f_{00} \\ *** & 0 & ***** & 1 & * & f_{01} \\ *** & 1 & ***** & 0 & * & f_{10} \\ *** & 1 & ***** & 1 & * & f_{11} \\ & | & \leftarrow \delta(H) \rightarrow & | & & \end{array}$$

The fitness values are schema averages, assumed to be constant with no variance (this last restriction may be lifted without changing our conclusions, as we only consider expected performance). With this problem definition, there are two ways to interpret the MDP. First, the problem may be taken literally — the $*$'s really don't matter and all strings with the specified two-bit similarity receive the specified fitness. More usually, however, we view each two-bit sub-problem as one of $\binom{l}{2}$ two-bit problems contained within the full l-bit problem. The latter orientation is the most useful; however, the former is easier to visualize and discuss so we will talk about the MDP as a single problem.

Thus far, we have introduced the two-problem, but we have not shown how we are going to make it deceptive, nor have we shown why the deceptive two-problem is the minimal, deceptive problem. To get at these things let us limit the problem somewhat. First of all, let's assume that f_{11} is the global optimum:

$$f_{11} > f_{00}; \qquad f_{11} > f_{01}; \qquad f_{11} > f_{10}.$$

Since the problem is invariant to rotation or reflection in Hamming two-space, the assumption of a particular global optimum is irrelevant to the generality of our conclusions.

Next, we introduce the element of deception necessary to make this a tough problem for a simple genetic algorithm. To do this, we want a problem where one or both of the sub-optimal, order-one schemata are better than the optimal, order-one schemata. Mathematically, we want one or both of the following conditions to hold:

$$f(0*) > f(1*)$$

$$f(*0) > f(*1).$$

In these expressions we have dropped consideration of all alleles other than the two defining positions, and the fitness expression implies an average over all strings contained within the the specified similarity subset. Thus we would like the following two expressions to hold:

$$\frac{f(00) + f(01)}{2} > \frac{f(10) + f(11)}{2}$$

$$\frac{f(00) + f(01)}{2} > \frac{f(01) + f(11)}{2}.$$

Unfortunately, we cannot have both expressions hold simultaneously in the two-problem (if they did, point 11 could not be the global optimum), and without loss of generality we assume that the first expression is true. Thus, the deceptive two-problem is specified by the globality condition (f_{11} is the best) and one deception condition (we choose $f(0*) > f(1*)$).

To put the problem into closer perspective we normalize all fitnesses with respect to the fitness of the complement of the global optimum as follows:

$$r = \frac{f_{11}}{f_{00}}; \qquad c = \frac{f_{01}}{f_{00}}; \qquad c' = \frac{f_{10}}{f_{00}}.$$

We may rewrite the globality condition in normalized form:

$$r > c; \qquad r > 1; \qquad r > c'.$$

We may also rewrite the deception condition in normalized form:

$$r < 1 + c - c'.$$

From these conditions, we may conclude a number of interesting facts:

$$c' < 1; \qquad c' < c.$$

From these, we recognize that there are two types of deceptive two-problem:

$$\text{TYPE I: } f_{01} > f_{00} \qquad (c > 1)$$

$$\text{TYPE II: } f_{00} > f_{01} \qquad (c \leq 1).$$

Figures 6-1 and 6-2 display sketches of representatives of these problems where the fitness is graphed as a function of two Boolean variables. We note that both cases are

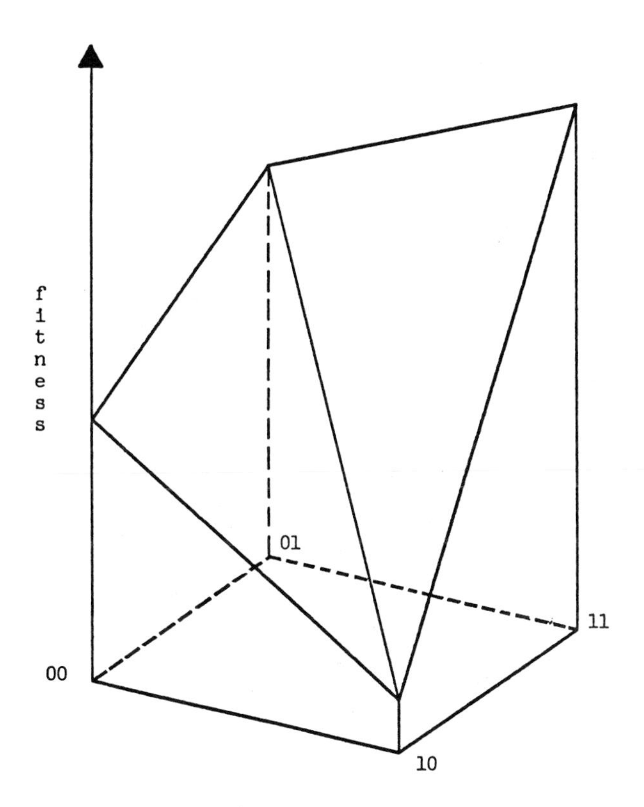

Figure 6-1: Sketch of type I, minimal, deceptive problem (MDP) $f_{01} > f_{00}$

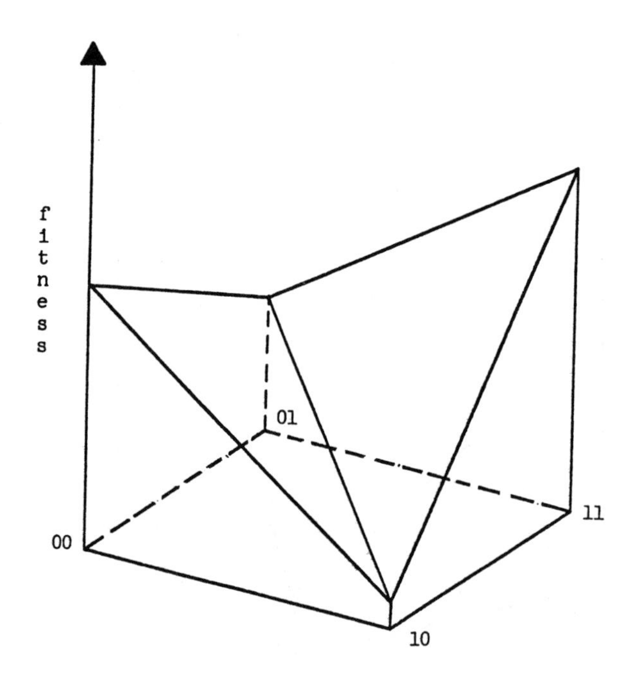

Figure 6-2: Sketch of type II, minimal, deceptive problem (MDP) $f_{00} > f_{01}$

X	00	01	10	11
00	S	S	S	01 10
01	S	S	00 11	S
10	S	00 11	S	S
11	01 10	S	S	S

Figure 6-3: Crossover Yield Table

deceptive, and it may be shown that neither case can be expressed as a linear combination of the individual allele values, since neither case can be expressed in the form

$$f(x_1 x_2) = b + \sum_{i=1}^{2} a_i x_i.$$

In the biologist's terms we have an epistatic problem. Since it similarly may be proved that no one-bit problem can be deceptive, the deceptive, two-problem is the smallest possible deceptive problem: it is *the* minimal, deceptive problem (MDP). With the MDP defined, we now turn toward a complete schema analysis of a simple genetic algorithm's behavior in solving the MDP.

6.4 Extended Schema Analysis Of The Two-Problem

Thus far we have constructed a generalized two-bit problem which seems capable of misleading a genetic algorithm when the two defining bits of the problem become widely separated on the string. Judging from the schema theorem we expect to have difficulty when the factor

$$\frac{f(11)}{\bar{f}}\left[1 - p_c \frac{\delta(11)}{l-1}\right]$$

is less than or equal to one (assuming that $p_m = 0$). A more careful analysis requires us to consider the details of crossover more closely.

The usual calculation of the expected number of schemata in the next generation is a lower bound. This is so, because the derivation contains no source terms and it assumes that we lose the schema whenever a cross occurs between the schema's outermost defining bits. In the two-problem, the mating and crossing of non-complementary pairs conserves the genetic material of the parents. For example, 00 crossed with 01 yields 01 and 00. The only time a loss of genetic material occurs is when complements mate and cross. In these cases, a 00 mated and crossed with a 11 yields the pair 01 and 10, and likewise, a 01 mated and crossed with a 10 yields the pair 11 and 00. The full crossover yield table is shown in Figure 6-3, where an S is used to indicate that the offspring are the same as their parents.

In the yield table we see how complements do lose material; although we also see how this loss shows up as a gain to the other complementary pair of schemata. Using this information, it is possible to write more accurate difference relationships for the expected proportion P of each of the four competing schemata. To do this we must account for the correct expected loss and gain of schemata due to crossover. Assuming proportionate reproduction, simple crossover, and random mating of the products of reproduction, we obtain the following autonomous, nonlinear, difference equations:

$$P_{11}^{t+1} = P_{11}^{t}\frac{f_{11}}{\overline{f}}\left[1 - p'_c\frac{f_{00}}{\overline{f}}P_{00}^{t}\right] + p'_c\frac{f_{10}f_{01}}{\overline{f}^2}P_{01}^{t}P_{10}^{t}$$

$$P_{10}^{t+1} = P_{10}^{t}\frac{f_{10}}{\overline{f}}\left[1 - p'_c\frac{f_{01}}{\overline{f}}P_{01}^{t}\right] + p'_c\frac{f_{00}f_{11}}{\overline{f}^2}P_{00}^{t}P_{11}^{t}$$

$$P_{01}^{t+1} = P_{01}^{t}\frac{f_{01}}{\overline{f}}\left[1 - p'_c\frac{f_{10}}{\overline{f}}P_{10}^{t}\right] + p'_c\frac{f_{00}f_{11}}{\overline{f}^2}P_{00}^{t}P_{11}^{t}$$

$$P_{00}^{t+1} = P_{00}^{t}\frac{f_{00}}{\overline{f}}\left[1 - p'_c\frac{f_{11}}{\overline{f}}P_{11}^{t}\right] + p'_c\frac{f_{01}f_{10}}{\overline{f}^2}P_{01}^{t}P_{10}^{t}$$

In these equations, the superscripts are time indices, and the subscript binary numbers are schema indices. The variable $\overline{f}$ is simply the current (generation t) population average fitness which may be evaluated as follows:

$$\overline{f} = P_{00}^{t}f_{00} + P_{01}^{t}f_{01} + P_{10}^{t}f_{10} + P_{11}^{t}f_{11}.$$

The parameter p'_c is the probability of having a cross and having that cross fall between the two defining bits of the schema:

$$p'_c = p_c\frac{\delta(H)}{l-1}.$$

Together, these equations predict the expected proportions of the four schemata in the next generation. With specified initial proportions, we may follow the trajectory of the expected proportions through succeeding generations. A necessary condition (albeit a weak one) for the convergence of the GA is that the expected proportion of optimal schemata must go to unity in the limit as generations continue:

$$\lim_{t\to\infty} P_{11}^{t} \to 1.$$

To examine the behavior of these equations more carefully, we look at several numerical solutions of the extended schema equations for type I and II problems. We also briefly present some theoretical results without proof.

6.5 Type I MDP Results

Suppose we have an MDP with the following fitness values:

$$f_{11} = 1.1; \qquad f_{01} = 1.05; \qquad f_{00} = 1.0; \qquad f_{10} = 0.0.$$

Thus, by previous definition, $r = 1.1$, $c = 1.05$, and $c' = 0.0$. The problem is type I because $c > 1$, and furthermore the problem is deceptive because

$$r = 1.1 < 1 + c - c' = 2.05.$$

To make the problem as hard as possible, we assume that $p'_c = 1$. In other words we are assuming that the two defining bits are at opposite ends of the string and that crossover is performed at every opportunity.

The extended schema equations have been coded in Pascal. Numerical results over 100 generations with equal initial proportions ($P^0_{ij} = \frac{1}{4}$) are presented in Figure 6-4. At first, the deception of the problem fools the genetic algorithm as the schema 01 (second best) increases and the schema 11 (the best) decreases; however, as the two schemata 00 and 10 die off, schemata 11 and 10 can only battle each other with the better of the two ultimately winning in the end.

This result may be generalized to theoremhood: any type I MDP converges (in the sense stated earlier) to the global optimum as long as there is some initial proportion of 11 schemata in the population. We will not give the complete proof here; however, the proof follows the performance observed in our example above. To prove the theorem, we simply show that one-schemata 1* and 0* battle for control (order 1 schema equations contain no source or sink terms) with $P(0*)$ going to 0 in the limit. Thereafter, the constituents of schema 1*, 11 and 10, battle it out with the stakes raised by the increased population average $\overline{f}$. Since 11 is the global best, it must win and eventually take over the entire population.

6.6 Type II MDP Results

The situation is somewhat the same for type II problems; however, the problem is no longer convergent for all initial conditions, although it is provably convergent for many probable initial conditions. To see this, we consider some numerical results and outline some sufficient conditions for convergence.

In our numerical experiments, let's look at a problem with the following fitness values:

$$f_{11} = 1.1; \qquad f_{00} = 1.0; \qquad f_{01} = 0.9; \qquad f_{10} = 0.5.$$

Thus, by definition $r = 1.1$, $c = 0.9$, and $c' = 0.5$. The problem is deceptive because the 0* schema is better than the 1* schema:

$$r = 1.1 < 1 + c - c' = 1.5.$$

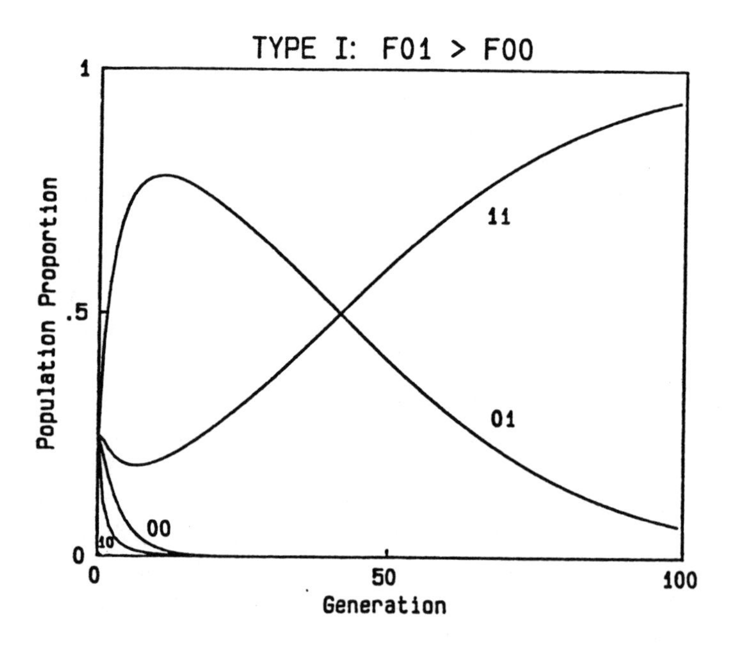

Figure 6-4: Numerical solution of a type I, minimal, deceptive problem (MDP) $r = 1.1$, $c = 0.9$, $c' = 0.0$

Once again we assume $p'_c = 1.0$ and calculate the expected proportion of each schema in succeeding generations. Figure 6-5 shows a numerical solution starting from the initial condition, $P_{ij}^0 = 0.25$. As in the Type I problem, the schema 01 gets an early foothold, but is unable to keep pace once the *0 representatives die out. On the other hand, if the 00 schema receives too great a proportion of the initial population, it can take over and crowd out the best (11). An example of this is shown in the numerical results of Figure 6-6. Here we use the same type II problem described above with different starting conditions:

$$P_{00}^0 = 0.7; \qquad P_{01}^0 = P_{10}^0 = P_{11}^0 = 0.1.$$

In this solution, the overwhelming preponderance of 00 schemata causes the GA to diverge as shown. A sufficient condition for convergence in the type II problem may be derived:

$$P_{11}^0 > \frac{f_{00}}{f_{11}} P_{00}^0.$$

The proof follows the argument for the type I problem with the added condition that the proportion of the complement never be allowed to get ahead of the proportion of the optimum.

6.7 Summary And Conclusions

In this paper we have reviewed the theory of schemata underlying GA operation, constructed the minimal, deceptive problem (MDP) — the smallest problem which may give a simple genetic algorithm difficulty — and examined some theoretical and numerical results from an extended schema analysis of the MDP.

It is a surprising, new result, that this order-two, deceptive problem does not cause a simple genetic algorithm to diverge uniformly under unfavorable conditions (high crossover rate, long schema defining length). On the contrary, it has been shown that under a wide range of starting conditions, the genetic algorithm converges (in expected proportion) to the global optimum regardless of crossover rate, and regardless of schema defining length. This result has not and cannot be obtained using the normal bounding, schema analysis. An extended analysis of the propagation of schemata has been developed which includes the correct crossover source and loss terms. The analysis of these nonlinear difference equations has identified theoretical limits on the convergence of a simple GA on the MDP.

These methods may be extended to higher order, deceptive problems. A three-problem may be tackled directly using the extended schema analysis technique to determine appropriate source terms and calculate the improvement in crossover survival probability. Many of the deceptive arrangements within the three-problem will, however, be nothing more than two-problems. This will reduce the number of cases that require examination. How far and whether this method can be extended to even higher order

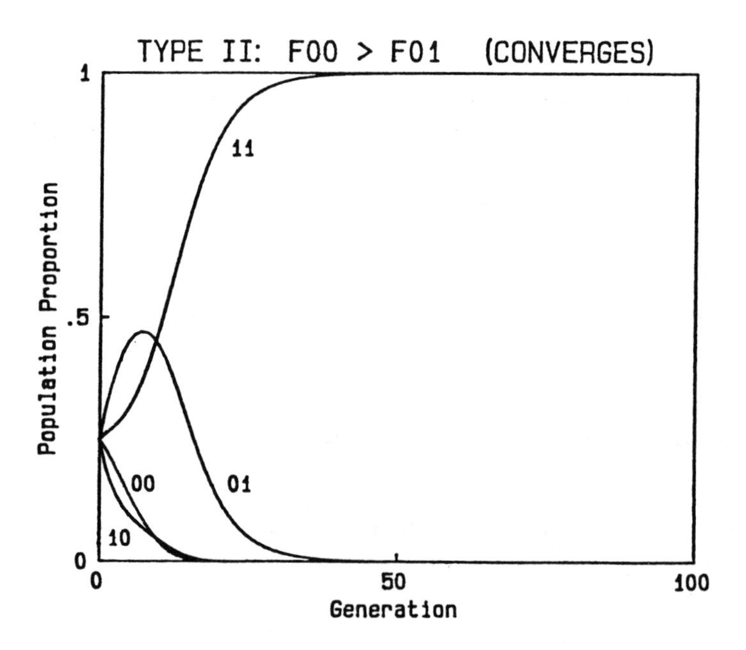

Figure 6-5: Numerical solution of type II, minimal, deceptive problem (converges) $r = 1.1$, $c = 0.9$, $c' = 0.5$, equal initial proportions

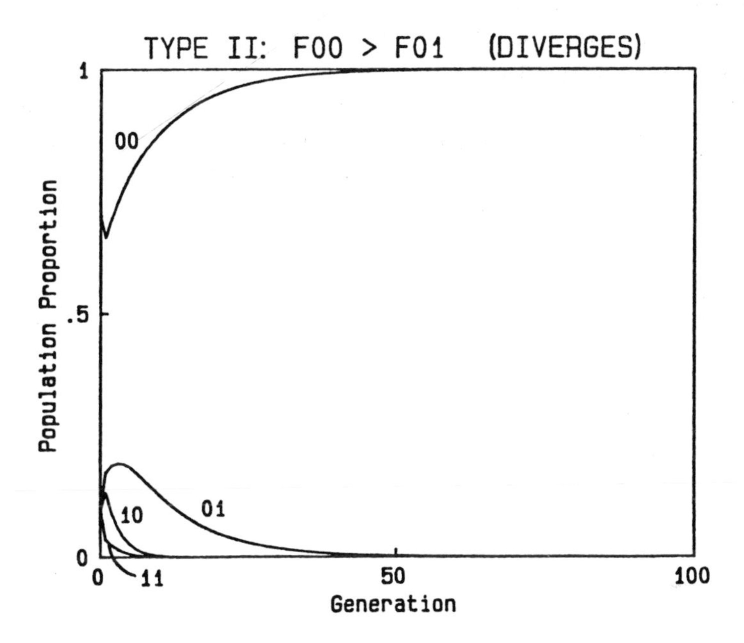

Figure 6-6: Numerical solution of type II, minimal, deceptive problem (diverges) $r = 1.1$, $c = 0.9$, $c = 0.5$, unequal initial proportions

problems is not yet clear; however, if the technique can be extended to such problems, even approximately, it will give genetic algorithm researchers a method of describing the class of GA-hard problems.

These results also have ramifications for the use and design of reordering operators like inversion. In one sense, these results tend to support the popular notion that reordering operators are not necessary in many GA applications. If all order-two schemata are processed intelligently, regardless of defining length and crossover rate, and if many problems are no more than order-two deceptive, then a simple GA is theoretically capable of finding good results without directly searching for better rearrangements of the string codings. On the other hand, for problems that are more than order-two deceptive, the type of analysis presented in this paper may be useful for calculating waiting times to acceptable rearrangements of the strings (under different reordering operators) which thereafter permit the optimum to be found.

J. David Schaffer

Chapter 7
Some Effects of Selection Procedures on Hyperplane Sampling by Genetic Algorithms

Abstract

This paper is concerned with the property of genetic algorithms called intrinsic parallelism. The proof of this property was developed by John Holland for a particular reproductive plan, but since then, a number of empirical studies have been conducted using variants of this procedure without theoretical consideration of whether these procedures still should exhibit this property. In this paper, two variant procedures are examined, differing from Holland's in the method of parent selection. Their effects on intrinsic parallelism are analyzed.

It is shown that these changes should significantly alter the hyperplane sampling, but that intrinsic parallelism is preserved. Furthermore, direct empirical evidence is presented showing intrinsic parallelism in action.

7.1 Introduction

In devising the class of computational procedures known as genetic algorithms (GA's), Holland (Holland 1975) provided a proof for an exciting property called intrinsic parallelism. This property says that a genetic algorithm will tend to sample from a space of strings in such a way that schemata (or more precisely, hyperplanes) from this space are represented in succeeding generations in proportion to their observed fitness relative to the population average. Moreover, this highly desirable sampling behavior occurs without the burden of having to compute the fitness averages for any of the hyperplanes. Since there are very many more than N hyperplanes represented in a population of N strings, this constitutes the only known example of combinatorial explosion working to advantage instead of disadvantage.

Since these concepts were first published, a number of empirical studies have been conducted to explore the behavior of these algorithms and to learn for what classes of computational tasks they might be best suited. (For some examples, see the proceedings of a recent conference (Grefenstette 1985).) Unfortunately, many of these implementations differ in some way from the GA upon which the proof of intrinsic parallelism rests (which I shall call the canonical GA) and analytical demonstration that these modified

GA's still exhibit this property are usually lacking. On the positive side, the empirical results seem generally to support the assumption that they do. Nevertheless, an analytical inspection of some of these procedures is in order.

After reviewing Holland's canonical GA and the proof for intrinsic parallelism, two variants will be considered. These differ from the canonical GA in the selection step only. In the canonical GA, one parent is selected on the basis of fitness while its mate is selected with a uniform distribution (*i.e.* all population members have an equal chance of mating with the good performer). First, a degenerate procedure in which neither parent is selected for fitness is examined to highlight the synergy between the crossover operator and the survival-of-the-fittest selection. Next, the selection of both parents for fitness is examined.

7.2 Holland Revisited

We begin by reviewing Holland's theorem for intrinsic parallelism. Figure 7-1 illustrates the canonical GA whose behavior will be analyzed. It is assumed that the reader is familiar with the crossover operator. Several good descriptions are available (see Booker's paper in this volume, as well as Holland 1975, DeJong 1980, and Grefenstette 1984, to name a few).

The generations referred to in Figure 7-1 are constant-size sets of fixed-length bit strings[1]. The strings will be called chromosomes. For chromosomes of length L, there is a set of schemata called hyperplanes which are also strings of length L, but are defined on the ternary alphabet $\{0, 1, \#\}$, where # represents "don't care". Hence, there are 3^L hyperplanes in this set, each of which defines a different subset from the space of chromosomes. In addition, each chromosome will be a member of 2^L of these hyperplanes. Therefore, the number of hyperplanes sampled in a generation of N chromosomes will be less than $N2^L$. Even though the actual number of hyperplanes represented in a given generation will be considerably less than this upper bound, it will be much greater than the number of chromosomes (N).

The reason for concentrating on these hyperplanes is that the fundamental theorem of intrinsic parallelism is a theorem on the sampling from this large space of hyperplanes achieved by the canonical GA operating on the much smaller set of chromosomes. This theorem says, in essence, that each hyperplane present in a generation will be sampled in the next generation at a rate proportional to its observed fitness relative to the

[1]There is nothing sacred about a binary character set. In principle any character set may be used. However, there are reasons for preferring binary. The arguments have been given by Holland (Holland 1975, p. 71) and by Smith (Smith 1980, p. 56) and some empirical evidence for the correctness of these arguments has been presented by Schaffer (Schaffer 1984, p. 107).

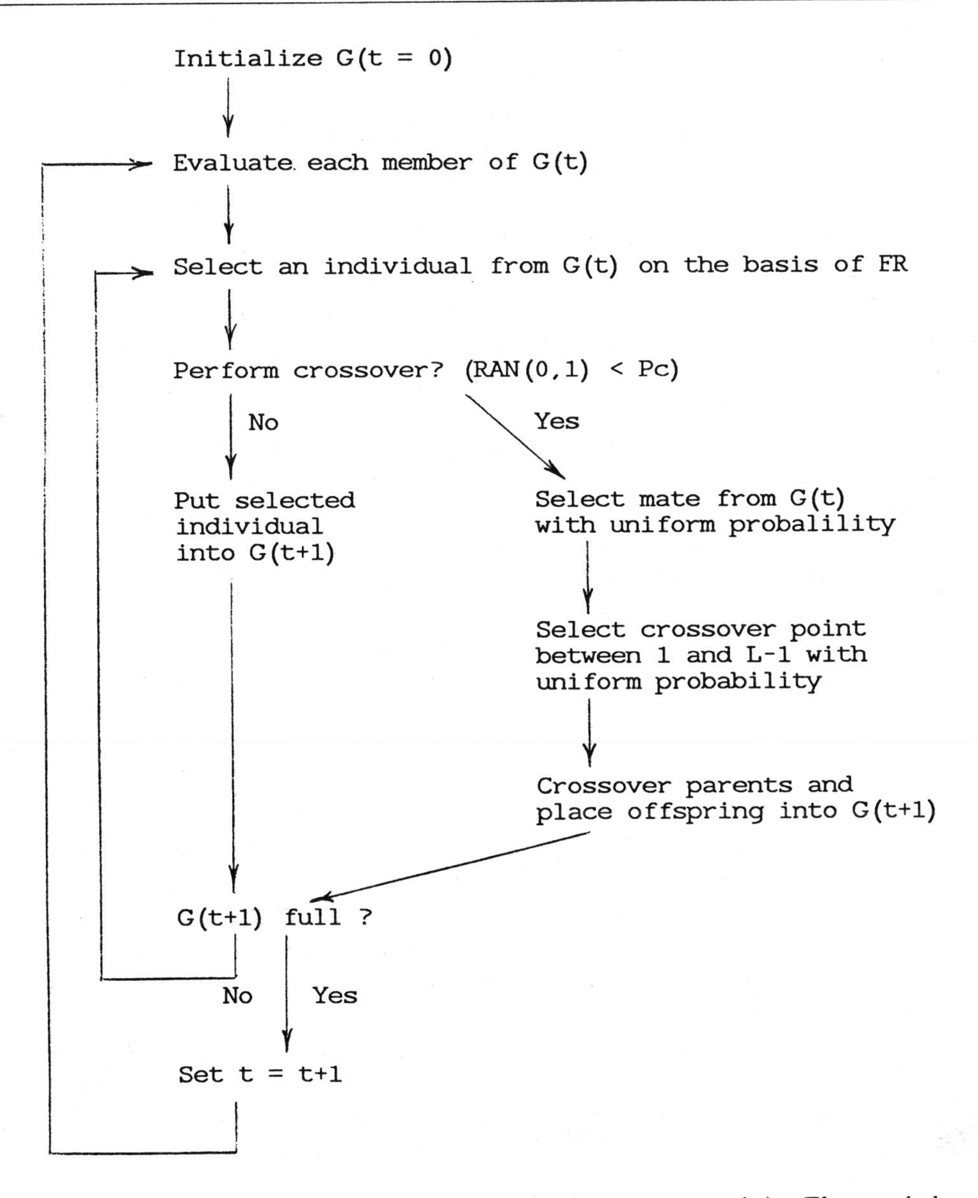

Figure 7-1: The canonical genetic algorithm (with crossover only). The symbols are explained in the text.

population average. More formally

$$P(\xi,t+1) \geq \left[1 - P_c \frac{l(\xi)}{L-1}(1 - P(\xi,t))\right] FR\ P(\xi,t) \tag{7.1}$$

where ξ designates a particular hyperplane and $P(\xi,t)$ designates the fraction of the population at time t that are members of hyperplane ξ. FR designates the fitness ratio (*i.e.* the ratio of the average fitness of the members of ξ to the population average. Strictly, FR should be given as $FR(\xi,t)$, but the arguments have been dropped for simplicity.) The defining length of a hyperplane, $l(\xi)$, is the length of the string from the first to the last locus that is not # and it varies from zero (for a hyperplane defined on only a single locus) to $L-1$ (for a hyperplane defined on at least the first and last loci of the chromosome). It is the number of places that the crossover point can fall within the hyperplane. P_c designates the crossover rate which is the probability that crossover will occur each time an individual is selected for reproduction.

The anatomy of this theorem is as follows. The product of the two terms outside the square brackets ($FR\ P(\xi,t)$) is the expected number of offspring from the members of ξ. The term in the square brackets is the expected proportion of these offspring who will not have their membership in ξ disrupted by the crossover operation and will therefore remain members of ξ in the next generation. This is simply one minus the probability of disruption. The probability of disruption is the product of the probabilities of three independent events, that crossover will occur (P_c), that it will fall within the defining length of the hyperplane ($\frac{l(\xi)}{L-1}$) and that the mate of the parent who is a member of ξ is not also a member of $\xi(1 - P(\xi,t))$. This last term comes from the selection of mates with a uniform probability and the fact that a crossover event between two members of ξ cannot fail to produce offspring who are members of ξ. Finally it is noted that the representation of ξ in the next generation is not limited to the offspring of the members in the current generation. Crossover between parents who are not members may produce offspring who are. Hence, the greater-than-or-equal-to relation.

To better understand the implications of this theorem, we will consider two extreme cases. The term $P_c\frac{l(\xi)}{L-1}$ represents the probability that a given hyperplane, present in one parent, will be disrupted. At one extreme, this term will be zero if either there is no crossover ($P_c = 0$) or the hyperplane is defined on only a single locus ($l(\xi) = 0$). In this case, the hyperplane cannot be disrupted and sampling reduces to the expected number of offspring.

$$P(\xi,t+1) \geq FR\ P(\xi,t) \tag{7.2}$$

At the other extreme, disruption is guaranteed if both $P_c = 1$ and $l(\xi) = L-1$. In this case equation (1) becomes

$$P(\xi,t+1) \geq FR\ P(\xi,t)^2 \tag{7.3}$$

Figure 7-2 shows plots of equations (2) and (3) for three values of the fitness ratio, 0.5, 1.0 and 2.0 (*i.e.* for hyperplanes that perform at half the population average, the average and

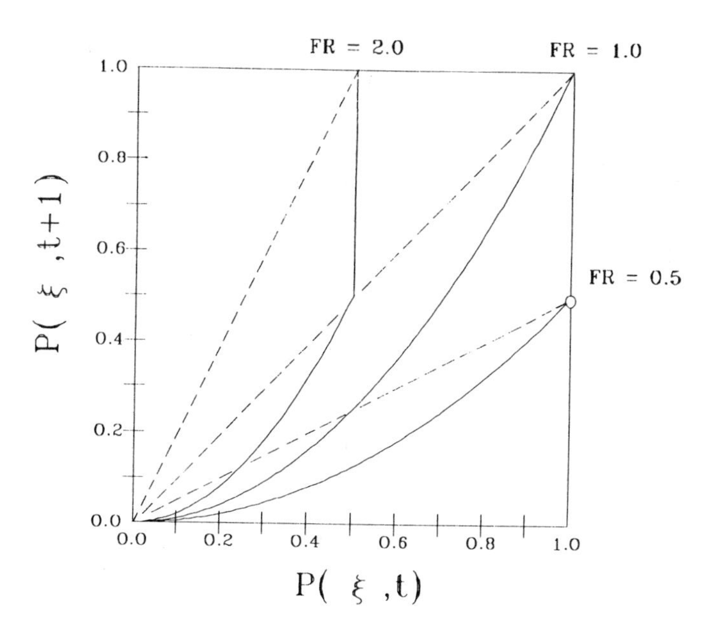

Figure 7-2: The expected sampling of hyperplanes performing at half, twice and exactly the population average under maximal (solid lines) and minimal (dashed lines) disruption by crossover.

twice the average). Two peculiarities of this figure are worth noting. When $P(\xi,t) = 1$, every member of the population is a member of the hyperplane and the hyperplane FR must also be one. Hence, the absent point for the $FR = 0.5$ curves. In addition, the definitions impose the constraint that the product of FR and $P(\xi,t)$ never exceed one. Hence, the $FR = 2.0$ curves must halt at $P(\xi,t) = 0.5$.

The difference between the minimal and maximal curves might be thought of as a measure of the bias shown by the canonical GA against hyperplanes with long defining lengths. It is more difficult to evolve a coadapted pair of genes if they are far apart on the chromosome because they are more likely to be separated by the crossover operator. Notice, however, that this bias is worse for the better than average hyperplanes than it is for the inferior ones. For example, a long hyperplane that is twice as good as the average will be expected to lose market share (*i.e.* its fractional representation in the population) at time $t+1$ if it has less than half the market at time t!

Let us examine this bias more closely. If we define the bias B as the difference between the expectations for the minimal and maximal disruption cases

$$B = FR\ P(\xi,t) - FR\ P(\xi,t)^2 \tag{7.4}$$

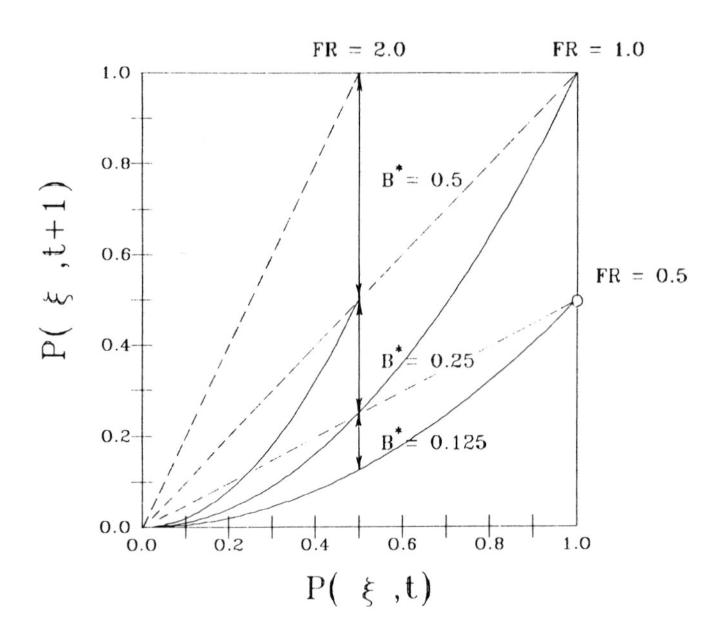

Figure 7-3: The relation of maximum bias to FR and $P(\xi, t)$ for the canonical GA

it reaches its maximum at

$$\begin{aligned} P^*(\xi, t) &= 0.5 \quad \text{for } FR \leq 2.0 \\ &= \tfrac{1}{FR} \quad \text{for } FR > 2.0 \end{aligned} \tag{7.5}$$

Furthermore, substituting for $P(\xi, t)$ in equation (4) yields

$$\begin{aligned} B^* &= 0.25FR \quad \text{for } FR \leq 2.0 \\ &= 1 - \tfrac{1}{FR} \quad \text{for } FR > 2.0 \end{aligned} \tag{7.6}$$

Thus the bias becomes worse with increasing FR. These relationships are illustrated in Figure 7-3.

7.3 Variation 1

Before considering a more practical variation of the canonical GA, it might be instructive to consider a simple degenerate case. What happens if we perform crossover without selection of fitness (*i.e.* random selection of both parents with a uniform distribution)? This amounts to setting FR to one in equations (2) and (3) and corresponds to the central curve pair of Figure 7-2. Under these conditions, no hyperplane is expected to gain market share[2]. Such a procedure would perform too much exploration and too little exploitation. Thus, the crossover operator must be combined with survival-of-the-fittest selection to work its magic.

7.4 Variation 2

The bias in the hyperplane sampling can be significantly altered by a simple expedient. By selecting both parents on the basis of fitness instead of only one, equation (1) becomes

$$P(\xi,t+1) \geq \left[1 - P_c\frac{l(\xi)}{L-1}\right] FR\ P(\xi,t) + P_c\frac{l(\xi)}{L-1}FR^2P(\xi,t)^2 \tag{7.7}$$

To see this, replace the $P(\xi,t)$ term within the square brackets of equation (1) by $FR\ P(\xi,t)$. The minimum disruption equation (2) remains unchanged, but the maximum disruption equation (3) now becomes

$$P(\xi,t+1) \geq (FR\ P(\xi,t))^2 \tag{7.8}$$

The corresponding plots are given in Figure 7-4.

Notice that the sampling bias has become better for the above average and worse for the below average hyperplanes. The break-even market share for long hyperplanes performing at twice the population average has dropped from 0.5 to 0.25. This makes intuitive sense because now both parents are more likely to be members of the above average hyperplanes and correspondingly less likely to represent below average ones.

Applying the same considerations of the bias to these equations yields

$$B = FR\ P(\xi,t) - FR^2P(\xi,t)^2 \tag{7.9}$$

for which

$$\begin{aligned} P^*(\xi,t) &= \tfrac{1}{2FR} \quad \text{for} FR \geq 0.5 \\ &= 1 \quad \text{for } FR < 0.5 \end{aligned} \tag{7.10}$$

and

$$\begin{aligned} B^* &= .25 \quad \text{for } FR \geq 0.5 \\ &= FR - FR^2 \quad \text{for } FR < 0.5. \end{aligned} \tag{7.11}$$

Thus, the maximum size of the bias has become independent of FR (above 0.5) and occurs at lower values of market share as FR increases. These relationships are illustrated in Figure 7-5.

[2]Except for those which might be introduced by crossovers between parents who are not members.

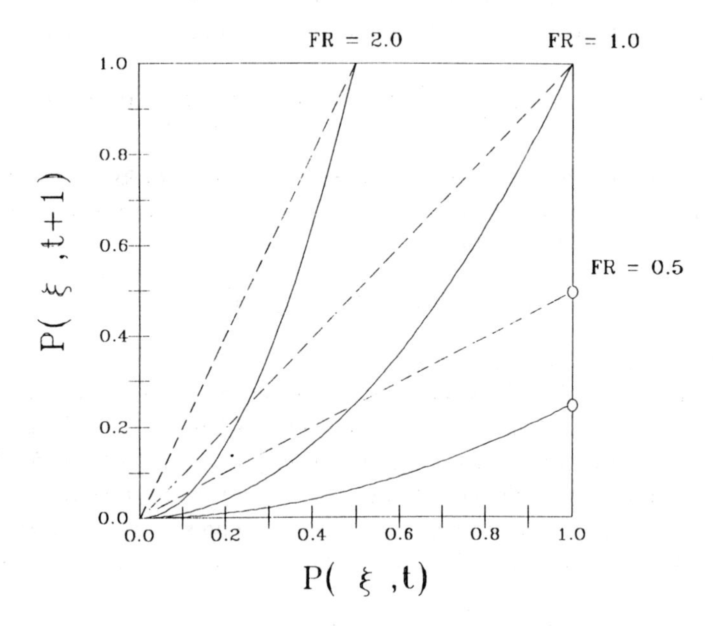

Figure 7-4: The expected sampling of hyperplanes performing at half, twice and exactly the population average under maximal (solid lines) and minimal (dashed lines) disruption by crossover when both parents are elected for fitness.

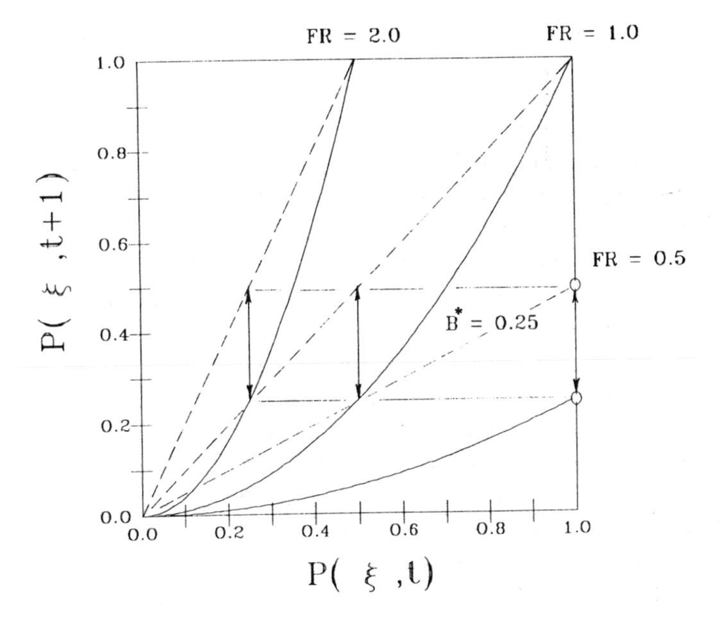

Figure 7-5: The relation of maximum bias to FR and $P(\xi, t)$ for GA variation 2

So much for theory, but does it work this way in practice? There are two aspects to this question. If we find different hyperplanes being sampled according to their observed fitness in a single run of a GA program, then we have an observation of intrinsic parallelism working. Furthermore, if we see hyperplanes with long defining lengths losing market share even when FR is above average, then we have an observation of the bias at work.

To test the theory, a GA was run to locate minima of a simple parabolic function of three variables:

$$f(x_1, x_2, x_3) = {x_1}^2 + {x_2}^2 + {x_3}^2. \tag{7.12}$$

The chromosomes were 30 bits long with 10 bits devoted to the binary coding of each x. A normal GA program (Schaffer 1984) was modified so that data could be collected on FR and sampling of specified hyperplanes. Note that FR is usually calculated for each individual in the population, but not for hyperplanes. The experiment was conducted with a population of 100 chromosomes, a crossover rate of 1.0 and a mutation rate of zero. The results for the first seven generations from a random initial population are shown in Figures 7-6–7-9 for four selected hyperplanes. The specific hyperplanes chosen were all sensitive to high-order bits of x_1 and/or x_3. For this function, hyperplanes sensitive only to low-order bits would be unlikely to perform consistently above or below the average and, therefore, would not provide clear illustrations. In Figures 7-6–7-9, each line segment shows how fitness ratio and market share changed from one generation to the next. The arrowheads show the progression of time from generation zero to generation seven. All data were from a single program run.

Figure 7-6 and Figure 7-7 show the expected fates of two hyperplanes with short defining lengths; the one above average increases its market share while the one below steadily loses. Figure 7-8 shows the fate of an above average hyperplane not present in the initial population. (The curve begins at the origin.) Upon first appearance in generation one, it is 40% better than average. It continues to gain market share until generation 7 where a retrograde step is observed. A more dramatic illustration of the demise of a better than average hyperplane can be seen in Figure 7-9. Even though almost 50% better than average in the initial population, this hyperplane has a long defining length and a small market share. It tends to lose market share and eventually (beyond generation 7) vanished from the population. Thus we see experimental evidence of intrinsic parallelism in action. Hyperplanes with short defining lengths are seen gaining or losing market share according to their fitness ratio, while simultaneously, hyperplanes with long defining lengths may fare poorly even if they continue to perform above the population average.

7.5 Conclusions

The sampling theorem that attributes intrinsic parallelism to GA's was derived for a particular instantiation from this class of reproductive plans. This canonical GA has been known to have a bias against proper sampling for hyperplanes with long defining lengths (Holland 1975). This paper has characterized this bias by showing that it is maximum when the hyperplane occupies 50% of the population for $FR \leq 2.0$ and at decreasing percentages for larger FR. Its magnitude grows with FR approaching one in the limit.

Furthermore, it was shown that a straightforward modification of the selection procedure could significantly alter, though not eliminate, this bias. By selecting both parents for fitness, an intuitively appealing thing to do, the magnitude of the bias is reduced to a constant (0.25) for all hyperplanes with $FR \geq 0.5$ and the population percentage at which it occurs is rendered an inverse function of FR. The cost for these gains is a worsening of the bias for hyperplanes with below average performance.

Then which GA is better? Unfortunately, this is less clear. By making takeover more difficult for above average hyperplanes, the canonical GA may tip the balance in favor of exploration at the expense of exploitation. For some task environments this may render the search more robust (*i.e.* less likely to converge prematurely). For other environments it may discard too many promising hyperplanes and miss good solutions altogether. On the other hand, the modified GA may produce better solutions quicker for some environments while prematurely converging for others. Unfortunately, few published papers on GA's give enough detail to know exactly what selection procedure was used so the empirical evidence is hard to weigh, but at least some experience with selecting both parents for fitness (Grefenstette 1984, Grefenstette 1986, Schaffer 1984) does not indicate a serious tendency to premature convergence.

Nevertheless, it is encouraging that variants of the canonical GA are possible without sacrificing intrinsic parallelism. The question of the best task domains for each of these variants remains to be researched.

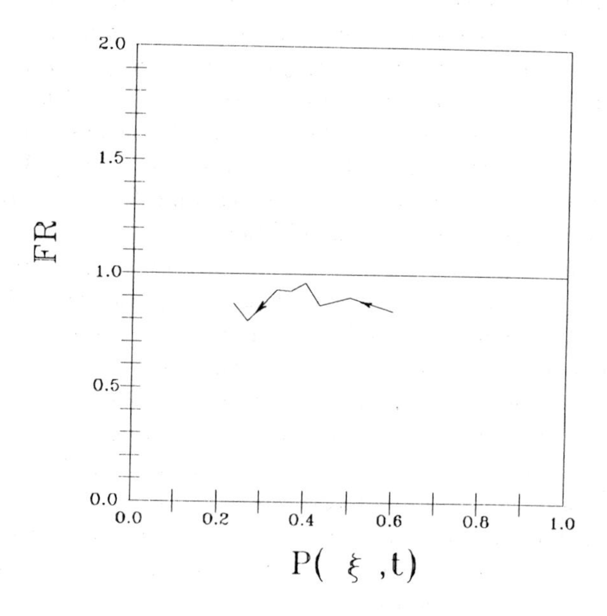

Figure 7-6: Hyperplane 1################################

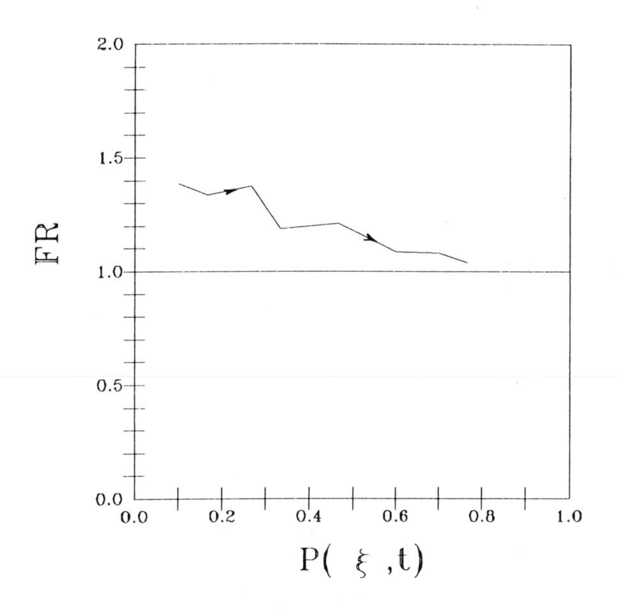

Figure 7-7: Hyperplane 011#############################

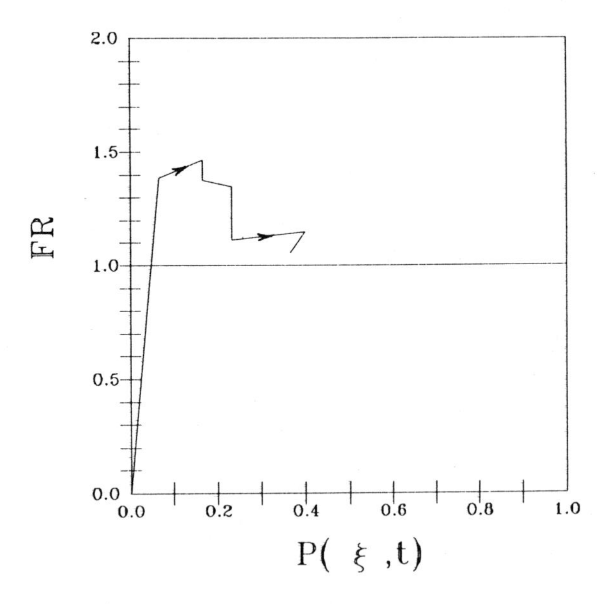

Figure 7-8: Hyperplane 01###################01########

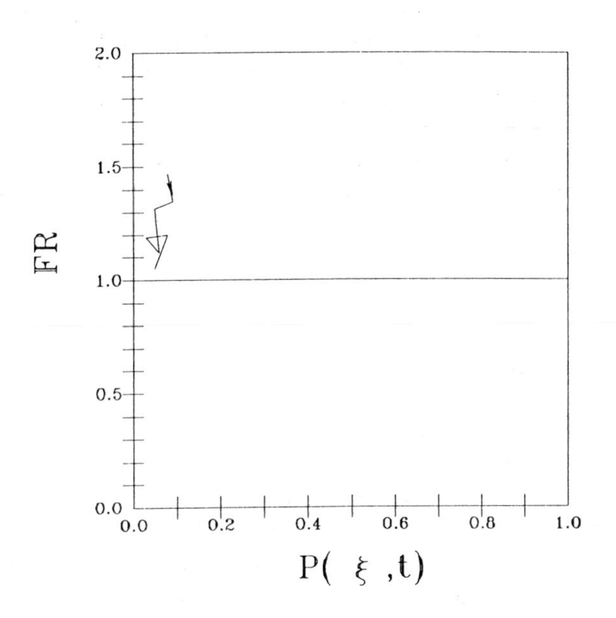

Figure 7-9: Hyperplane 1###################10########

Stewart W. Wilson

Chapter 8
Hierarchical Credit Allocation in a Classifier System

Abstract

Learning systems which engage in sequential activity face the problem of properly allocating credit to steps or actions which make possible later steps that result in environmental payoff. In the classifier systems studied by Holland and others, credit is allocated by means of a "bucket-brigade" algorithm through which, over time, environmental payoff in effect flows back to classifiers which take early, stage-setting actions. The algorithm has advantages of simplicity and locality, but may not adequately reinforce long action sequences. We suggest an alternative form for the algorithm and the system's operating principles designed to induce behavioral hierarchies in which modularity of the hierarchy would keep all bucket-brigade chains short, thus more reinforceable and more rapidly learned, but overall action sequences could be long.

8.1 Introduction

Many learning systems face the problem of temporal credit allocation: the proper reinforcement of activities which do not directly result in need satisfaction or external reward but are nevertheless essential precursors to such outcomes. Animals learn extensive hunting, stalking, or foraging behaviors aimed at the ultimate payoff of something to eat. A person who values others' cooperation must discover and reinforce effective precursor strategies. In the message-passing, rule-based *classifier systems* (Holland 1986), credit is allocated by means of a "bucket-brigade" algorithm to earlier-acting rules which "set the stage" for later actions that bring external payoff. The essential idea is that classifiers which match messages and become active on a given time step "pay" a fraction of their "strengths" to the strengths of the classifiers which posted the messages and were active on the previous time step. When finally external payoff enters the system, it is added to the strengths of the then currently active classifiers. If over time a given payoff-achieving sequence gets repeated, strength increments will in effect flow back to reinforce its early-acting classifiers. Consequently, early-acting classifiers that indeed participate in sequences that make possible later payoff will, by the algorithm, receive due credit.

In certain other AI systems which learn to perform multiple-step tasks [*e.g.*, Mitchell's LEX system for symbolic integration (Mitchell, Utgoff, and Banerji 1983), and the ACT* cognitive architecture of Anderson (1983)], credit is assigned to early steps by keeping and analysing a record of all pre-payoff actions, both considered and taken, and the associated reasoning. In contrast, Holland's bucket brigade technique does not depend on retrospective analysis but operates locally, during performance, in the strength transaction between steps, with the better classifiers at each step being selected statistically over time. The bucket-brigade principle would consequently appear appropriate for systems —such as animals and autonomous robots in on-going interaction with uncertain environments—where storage and analysis of raw experience is expensive or impractical.

In this paper, however, we suggest that the bucket-brigade may lose effectiveness as action sequences grow long. As a remedy, we propose a modification of the algorithm that makes it more directly reflect the hierarchical nature of behavior. We present first a brief description of classifier systems sufficient to understand the current algorithm. Then reasons are given to suggest that the bucket brigade may have trouble with longer sequences. In the main part of the paper we describe our "hierarchical bucket brigade" proposal.

8.2 Classifier Systems and the Bucket-Brigade Algorithm

The standard classifier system consists, structurally, of (1) a finite population [P] of fixed-length condition-action rules called classifiers, (2) a message list, (3) an input interface consisting of a set of environmental feature detectors, and (4) an output interface for affecting the environment. Sources of environmental payoff must also be defined. Functionally, the system employs three algorithms: a performance algorithm, a reinforcement algorithm (the bucket brigade), and a discovery algorithm.

The classifier population may be thought of as a set of hypotheses representing the system's current estimate of the best means of obtaining payoff. The performance algorithm, together with the classifier population, determines the system's short-term behavior. The reinforcement algorithm defines how payoffs received alter classifier strengths; that is, how credit is to be apportioned to the apparently responsible classifiers. The discovery algorithm, a background process, uses a version of Holland's (1975) genetic algorithm to generate new, possibly better classifiers by recombining "building blocks" from existing high-strength classifiers and inserting the new classifiers into the population to compete for payoff. [For reasons of space, we omit further description of the discovery algorithm since its operating details are not directly relevant to the present discussion. We shall assume that the algorithm is effective in introducing whatever classifiers are needed.]

The system's basic elements are *classifiers* and *messages*. All messages are strings of length L from $\{1,0\}$. All classifiers have the structure:

$$t_1, t_2, \ldots t_r / a$$

where the condition and action are separated by "/". The condition consists of r individual *taxa*, t_i; the action, a, is a message. Each taxon is a string of length L from $\{1, 0, \#\}$. The condition is satisfied if and only if every t_i matches some message currently on the message list. An individual t_i *matches* a message if and only if for every 1 or 0 in t_i the same value occurs at the corresponding position in the message; "#" functions as a "don't care" symbol in a taxon and matches unconditionally. The #'s confer generality on the taxon (and thus on the classifier) in the sense that 2^n distinct messages can be matched by a taxon containing n #'s. As just mentioned, each classifier also possesses a *strength*, a scalar quantity which is adjusted by the reinforcement algorithm and is intended to represent an estimate of the classifier's value to the system.

Messages from the environmental input interface have the same format as classifier action messages: they are strings of length L from $\{1, 0\}$. The output interface is a pre-defined mapping from some subset of the message space to actual actions in the environment. For example, the interface might respond to a message ending in "0110" by causing the system to take a step forward, etc.

With these details covered, we can understand the operation of the standard system. Time is discrete, divided into time-steps. On each time-step, the performance and reinforcement algorithms execute once, together forming a cycle which proceeds as follows:

1. Place all messages from the input interface on the current message list.

2. Compare all messages to all classifier taxa and form the *match set* [M] of classifiers with satisfied conditions.

3. Compute the *bid* B of each classifier C in [M] by taking the product of C's strength, its *specificity* (defined as the fraction of the positions in its taxa not filled by #'s), and a small constant (say 0.1).

4. Form the set [M′] consisting of the high bidders (using a threshold or some other well-defined procedure).

5. For each C in [M′], first reduce C's strength by the amount of its bid; then determine the set of classifiers which sent the messages which C matched, and increase their strengths by the amount of C's bid (the bid is shared among them in the simplest version).

6. Erase the current message list. Form a new message list from the action parts of the classifiers in [M′].

7. Process the new message list through the output interface (resolving effector conflicts by a well-defined procedure) to produce system output.

8. If payoff is received from the environment, divide it in equal shares among the classifiers of [M′], increasing each classifier's strength by the share amount.

9. Return to step 1.

In steps 1-4, classifiers which match the current situation (as represented by the message list, including environmental messages) compete for the right to post their messages on the next message list. The winners, forming the set [M′], are those with high strength (and/or specificity). In step 5, the bids of [M′]'s classifiers are in effect "paid" to the classifiers which belonged to [M′] on the previous cycle. At the same time, by placing their messages on the new message list in step 6, the members of the current [M′] gain the possibility of receiving payment from the classifiers of [M′] on the next cycle. The current [M′] may also receive payoff directly from the environment in step 8.

Steps 5, 6, and 8 implement the bucket-brigade algorithm. As noted earlier, the intention of the bucket brigade is to strengthen classifiers which, though they do not receive environmental payoff directly, nevertheless participate in sequences which lead to payoff. In time, if such sequences are repeated, environmental payoff will in effect flow back to the "stage-setting" classifiers. Those classifiers will gain strength and reach a new equilibrium where, in the transaction of step 5, they receive at least as much as they pay out. In contrast, other classifiers which are not coupled into payoff-achieving sequences will lose strength (because their messages are not matched on the next cycle or are matched only by weak classifiers) and, through step 4, they will lose their ability to influence system behavior.

8.3 Long Chains

The mechanics just described suggest that a classifier whose early action indeed contributes to later payoff may still have difficulty getting reinforced if the number of time-steps from its activation to payoff is large. As a simple example, suppose that classifier C posts a message which, through step 7 above, causes an action (*e.g.*, application of hand pressure to a restaurant door) that leads eventually, n time-steps (message-postings) later to payoff in the form of satisfaction at the taste of food. N will be of the order of the number of intervening elementary actions, which may be very large. If over time C is to be properly reinforced as a member of the sequence, the sequence will have to be repeated as many times as it takes the strength increment due to the food payoff to "reach" C.

By a simple simulation, we can estimate the number of repetitions required. Let a bucket-brigade sequence of length n be represented by a list of n elements, each initialized to a "strength" of zero. We represent one repetition of the whole sequence by having each element i carry out an analog of step 5 above. Specifically, the element's strength at time $t + 1$ is given by

$$S_i(t+1) = S_i(t) - eS_i(t) + eS_{i-1}(t),$$

except that the strength of the 0th element, since it receives payoff R from the environment and not from another element, is given by

$$S_0(t+1) = S_0(t) - eS_0(t) + R.$$

Each list element represents a classifier and its strength. The factor e stands for the product of specificity and the small constant mentioned in step 3, which we take to be the same for all classifiers in the chain. We also assume for simplicity that classifier $i-1$ matches only one message, namely that posted by classifer i, and that all the classifiers in the chain are different.

For convenience in the above equations we treat all the strength updates of one execution of the bucket-brigade sequence as occurring in a single unit of time. As the simulation runs, we observe a "wave" of strength flowing down the list. Our original question becomes: what is the value of t at which S_n reaches some criterial level? To answer this we note that as t increases, all strengths approach an asymptote equal to R/e, as is suggested by direct substitution (and can also be proved). Taking our criterial level to be 90% of R/e, the simulation gave, to a very close approximation,

$$t_{90\%} = \frac{3 + 1.2n}{e}$$

for values of e in the range from 0.1 to 0.4. E should be kept small so that classifier strengths average over a number of payoff events; typically, e is chosen to be no greater than 0.1. Given that value, our equation says that a "stage-setting" classifier just 10 steps from environmental payoff will require no fewer than 150 repetitions of the sequence to be properly reinforced.

But the situation is actually worse than this, since we have assumed that each activation of classifier n will be followed reliably, and in n steps, by external payoff to the system. This is not in general the case due both to stochastic elements which are usually built into steps 4 and 7 of the performance and reinforcement algorithm and to environmental surprises. Thus, even in the case of a "correct" sequence, the system may at some point occasionally go off onto a less remunerative path, resulting in net strength losses to the early-firing classifiers. From the analysis so far, it therefore appears that classifier systems will have difficulty learning complex behaviors (*i.e.*, those relying on extended action sequences).

8.4 Behavioral Modules

Clearly, something is wrong if the reinforcement algorithm must feed strength increments back through the enormous number of elementary steps between, say, the push on a restaurant door and the enjoyment of food. Intuitively, that sequence consists of just a few big steps: <enter restaurant>, <get a table>, <get food>, <eat>. If the algorithm treated *these* as the bucket-brigade units, reinforcement would be faster since the chain would be short. Somehow we must also reinforce the smaller steps which compose the big ones. But we note that <enter restaurant> can be broken into the sequence: <find door>, <open door>, <go through>, and that <open door> expands, in turn, into a short sequence one of whose components is <push>. Albus (1979), among others, shows how any complex activity can be decomposed into a hierarchy of behavioral modules each consisting of just a few "steps". If the bucket brigade could apply hierarchically to module steps, we might be able to reinforce quite extended activities without encountering the "long chain" problem.

Holland (1985) notes that long chains may be hard to reinforce under the bucket brigade, but suggests that "bridging" or "epoch-marking" classifiers would arise that propagate external payoff nearly directly to the early classifiers of the chain. Under the performance and reinforcement algorithm, the message list is replaced (step 6) on every cycle. But a certain classifier could be persistently (repetitively) active over an interval T if on each cycle within T its conditions were satisfied by some combination of unchanging features of the environment, its own message just posted in the last cycle, or a message from another persistent classifier. Consider now a classifier D which is separate from, but persistently active during the execution of a long bucket-brigade sequence. Suppose also that D matches the message sent by an early classifier of the sequence, and is also active when external payoff occurs. Holland points out that D would pass an increment of strength back to the early classifier on the *very next* repetition of the sequence, in effect short-circuiting the sequence and providing immediate "support" to an early member. Holland further suggests that persistent classifiers would naturally come to represent goals to be carried out by the sequences they bridge, and that hierarchies of such goals would arise automatically (*i.e.*, through selection pressure and the discovery algorithm). Thus Holland recognizes the bucket brigade's "long chain" problem, but believes that its solution requires little or no adjustment to the standard system.

Here we explore the alternative approach of modifying the performance and reinforcement algorithm so as explicitly to encourage behavioral modules and short bucket-brigade chains. The basic change is to use a hierarchical message list instead of the current homogeneous one in which all messages have equal status, and, for the moment, to allow at most one message at a time on a given level. Our plan of exposition is first to take the reader through an example, then to present the new algorithm, and finally to discuss questions which the algorithm raises.

8.5 An Example

Figure 8-1 illustrates the operation of the hierarchical performance and reinforcement algorithm over a certain interval of 17 time-steps. The figure shows principally the contents of the message list, but also indicates environment changes, actions, and bucket-brigade flows.

At time t_0 we imagine that the message M_1 spontaneously appears on a previously empty message list. M_1 is special in that it represents an internal system need, *e.g.*, <get food>, in which case we could say that the system has just felt (renewed) hunger. We note that M_1 stays on the list until the very end of the epoch, when food (R) is received. M_1 is in effect the *name* of a behavioral module (intent, plan, subprogram) with the purpose <get food>.

At t_0 the environmental input message was E_1 (top of figure). Since the system took no external action on that time step, the same environment holds (we assume) at t_1. But the overall situation is different at t_1 since the message list contains M_1. The system now forms its usual match set with the proviso that classifiers in the set must match both E_1 and M_1. From the match set, a *single* classifier is picked (based on a bid-like quantity) and that classifier's message, M_2, is posted on the list on the next level down. The interpretation is that M_2 names a module of M_1 that applies when the environment is E_1.

Still the system has not made an external action. At t_2, a match set is again formed with the proviso that its members must match E_1 (still unchanged) and M_2, illustrating the matching rule: "if no external action occurred in the previous time-step, compare only against the *lowest* level message on the list in forming the match set." The rationale is that the lowest level message represents the system's most immediate intent, which should have priority. Again, the figure shows the posting of message M_3 and thus a deepening of the hierarchy.

At t_3 something new happens. Following the same rules as above, the system picks a winning classifier whose message specifies an external action A_1. In this case the action is taken and no new message is posted (action messages cause only actions). We have reached the level of a module (M_3) whose components are not intents or submodules but external activity.

At t_4, a new matching rule applies: "if an external action occurred during the previous time-step, compare against *all* messages on the list; if the (again) *single* winning classifier matched a message on level k of the list, erase all lower-level messages (if any) and post the winner's message one level below k." In the current case, we see from the figure that M_3 must have been the highest message matched (since no messages got erased) and that the winner's message was the action A_2. The interpretation is that the system simply executes another action belonging to M_3.

At t_5, bigger changes occur. The second matching rule (the "ascent" rule) again applies, and this time the winning classifier matched M_2 (and E_3), resulting in erasure of M_3 and the posting of M_4. Here the interpretation is that, given environment E_3, module M_2 moves on to its second submodule M_4; *i.e.*, its first submodule, M_3, has been successfully carried out.

We now have enough information to understand the rest of the figure. From t_6

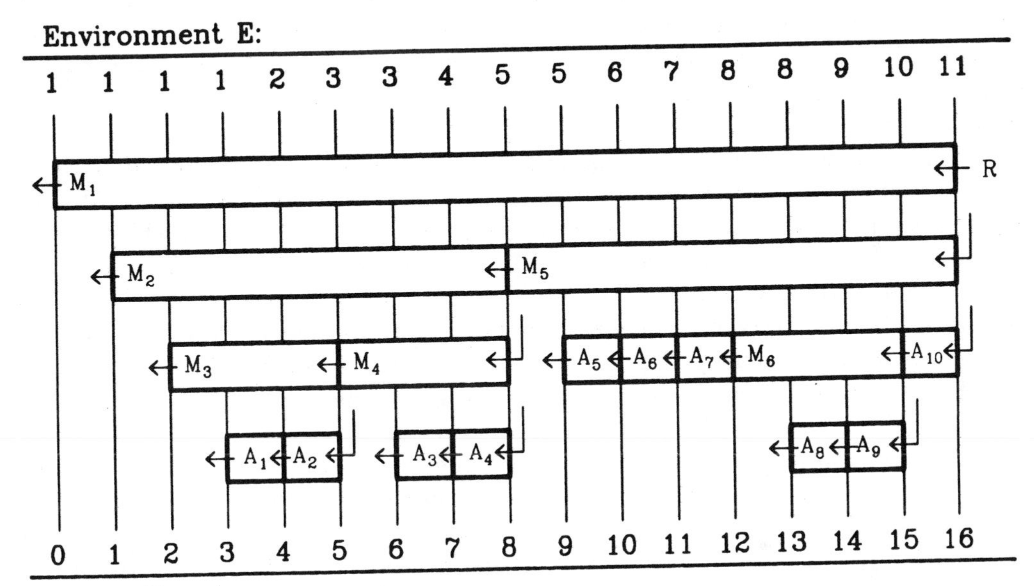

Figure 8-1: Example of Hierarchical Performance and Reinforcement Algorithm

through t_8, the system executes the actions of M_4, but this also completes M_2. At t_9 the system enters "descent" (the first matching rule applies) and begins execution of the module M_5, which lasts until t_{16}. Note that the first three steps of M_5 are actions but the fourth is a submodule. Finally, the fifth step, the action A_{10}, results in external reward entering the system, which causes erasure of the entire message list.

The "ascent" matching rule, which acts most dramatically at t_8 and t_{16}, is designed to identify the highest-level module to which the environment resulting from the current action is relevant, and to terminate all lower level modules. This corresponds to the observation that completion of a high-level subplan usually means completion of all subplans which immediately underlie it. For example, completion of the subplan <get a table> under the plan <get food> also completes <take a seat> and, under that, <pull the chair back up to the table>, etc.

The operation of the bucket brigade in Figure 8-1 is illustrated by the small arrows, which indicate strength flows. An arrow from one message to another, as between M_4 and M_3, means a payment from the classifier which posted M_4 to the classifier which posted M_3. As usual, the amount involved is a fraction of the strength of the source classifier, and it is added to the strength of the recipient. Similarly, an arrow from one action to another (or from a message to an action, or vice versa) means a payment between the two corresponding classifiers. The special case of an arrow leaving the first step of a module, as with M_2, means a fraction of the strength of the posting classifier is simply removed and "thrown away".

At time-steps 5, 8, 15, and 16, a more complicated payment pattern occurs. For example, at t_8, the standard strength fraction is deducted from the classifier which sent M_5, but the resulting quantity is paid to *each* of the three recipients indicated by the arrows. That is, if an amount Q is deducted from the source classifier, each recipient has its strength incremented by Q. Similarly, at t_{16}, the payoff quantity R (and not one-third of R) is paid to each of the three recipients shown. The intent of this "non-splitting" of payoff is to encourage hierarchical deepening where appropriate; a different rule may of course turn out to be better.

We may note in Figure 8-1 how the bucket-brigade pattern causes strength flows along the constituent steps of each module, thus reinforcing the steps in the spirit of the original bucket-brigade principle. But this "hierarchical bucket brigade" also achieves our objective of reducing the length of any individual chain. Note that the overall activity of Figure 8-1 consists of ten action steps (and 17 time-steps) yet no classifier is more than five payment steps from the external reward. More generally, hierarchy means that the average payment sequence length will be of the order of $\log n$, where n is the number of steps in the overall activity.

8.6 Hierarchical Performance and Reinforcement Algorithm

We now state the heirarchical performance and reinforcement algorithm, staying as close as possible to the form of the algorithm of Section 8.2.

1. Obtain the current message E from the environmental input interface.

2. If phase="descent", form the match set [M] of all classifiers which match both E and the lowest-level message on the message list, else

   If phase="ascent", form the match set [M] of all classifiers which match both E and any message on the message list.

3. Compute the bid B of each classifier C in [M] by taking the product of C's strength, its specificity, and a small constant (say 0.1).

4. Select a classifier C* from [M] using a procedure in which higher-bidding classifiers are more likely to be selected.

5. Reduce C*'s strength by the amount of its bid;
   then
   If phase="ascent", pay an amount equal to B to each of the classifiers (if any) which sent messages lower on the list than the message matched by C*, erase those lower messages, and pay an amount B to the classifier whose action was carried out on the previous time-step;

6. If C*'s message is an external action, set phase="ascent",
   Else post the message on the next lower empty level of the message list and set phase="descent".

7. If the message of step 6 was an action, take it.

8. If payoff R is received from the environment, pay amounts equal to R to each of the classifiers which sent messages now on the list, erase all messages, pay an amount R to the classifier whose action was just taken, and set phase="descent".

9. Return to step 1.

The new algorithm leaves some operational questions unanswered. For instance, we are not told what to do in step 2 if the match set [M] is null (this is also not covered in the standard algorithm). In "descent", the sensible thing would seem to be to assume that the most recent posting (lowest message) was a "mistake", erase it, and retry the match against the next higher message. In "ascent" the situation is more complicated, but failure to match is less likely since the whole list is matched against. A possible response would be to "reverse" the last action (if possible) and retry the match. In both cases, a stochastic element in the selection of C* (step 4) would permit alternative outcomes. If the system became truly stuck in a certain state or loop, a "fatigue" process could come into play, causing messages gradually to drop from the list. All these questions are more properly addressed at the level of the system routine of which the hierarchical algorithm is a component.

8.7 Discussion

An important difference between the hierarchical classifier system outlined here and the standard system is that parallelism appears to be greatly reduced. The standard system permits numerous messages to be posted in each cycle, whereas the hierarchical system permits the addition, to those already on the list, of no more than one message per cycle. Parallelism in the standard system is intended to serve several functions (Holland 1986). Having numerous classifiers active on each cycle should allow the system simultaneously to test numerous hypotheses about the best way to get to payoff. Over time, those that profit in the bucket brigade should win out (and, under the discovery algorithm, become progenitors of new, possibly even better, classifier hypotheses). Secondly, parallelism should give the system gracefulness in the sense that when control is divided among a cluster of rules, the failure or absence of one rule can be expected to have only a marginal effect on performance. Finally, complex situations may be more flexibly represented internally by a set of numerous activated rules, each responding to an element of the situation, than by a few, or just one, rule which would have to encompass all relevant aspects in its condition. In short, multiple activation is intended to give the system a more powerful *mental model* of the world (Holland *et al*, 1986).

This is clearly an important objective for any learning system. We suggest, however, that the hierarchical system is not so narrow as it may appear. At any moment, in general, the message list contains a number of messages, which could be taken to represent a mental model, but in this case an hierarchical one. The higher level messages represent broader, more general, aspects of the situation than the lower level messages. Selection of a message for posting on a given level is the result of a competition which on another occasion could well pick a different classifier's message for testing. Furthermore, the hierarchical system can be modified to permit more than one message on each level (essentially, one lets C* be a set instead of a single classifier, but there is not space here to go into detail). The single place where the hierarchical system is clearly "narrower" than the standard one is in "descent", where our rule is that only the lowest-level message gets matched against, corresponding to the principle that a plan cannot be achieved before its subplans. (To prevent insensitivity to environmental surprises, a multi-level interrupt can be provided by adding "If E differs from the previous E, set phase='ascent'" to step 1.)

Further research is needed to determine the hierarchical classifier system's merit.

We intend to apply it in an extension of our previous "animat", or artificial animal, simulations (Wilson 1985b, Wilson 1986b). In concluding, we would stress the hierarchical system's two apparent plusses: shorter bucket-brigade chains and explicit modularity.

John H. Holland, Keith J. Holyoak, Richard E. Nisbett, and Paul R. Thagard

Chapter 9
Classifier Systems, Q-Morphisms, and Induction

9.1 Introduction

This collection of excerpts from our recent book *Induction: Processes of Inference, Learning, and Discovery* is concerned with our motivation for using classifier systems – collections of pattern-matching bit strings with genetic search routines and a "bucket brigade" payoff mechanism – to model natural processes of induction[1]. In the sections that follow, we describe those features of inductive systems that we undertook to model, contrast the classifier system approach with other approaches to the induction problem, show formally with the notion of Q-morphisms how our use of bit strings allows compact encoding of knowledge in the form of abstraction hierarchies, and conclude by showing that the genetic algorithm is as effective at finding and bringing together useful combinations of genes in the induction domain as it is in the optimization domain.

9.2 Characteristics of Inductive Systems

Our framework is based on a set of guiding principles concerning the fundamental characteristics of inductive systems[2]. These principles motivated many of our decisions in designing the classifier system model of induction.

1. General knowledge can be represented by condition-action rules. These rules can vary enormously in the complexity of their conditions and actions, representing features that range from elementary perceptual ones to highly abstract categories. The immediate actions of rules consist of the posting of "messages" internal to the system.

2. Rules can represent both diachronic relations (for instance, between current and expected future states) and synchronic relations (associations and recategorizations of categories), and the two types of rules act together to generate inferences and

[1]This paper consists of excerpts from the book *Induction* (Holland, Holyoak, Nisbett, and Thagard 1986) selected by John Holland. Included also are annotations by John Holland that set these excerpts in context. The excerpted material appears by permission of The MIT Press.

[2]This section is taken from pp. 21–22 of *Induction*.

solutions to problems. Problem solving involves both diachronic search and synchronic recategorization of elements.

3. Higher-order knowledge structures such as categories correspond to implicit or explicit clusters of rules with similar conditions. Larger structures are thus composed of more elementary building blocks. (A realistic system must have inductive mechanisms for constructing these larger structures, a notion we refer to as the "principle of inductive adequacy".)

4. Superordinate relations among categories and rules yield an emergent default hierarchy. Exceptional information about specific examples will tend to override default rules, with the consequence that imperfect default rules will be protected from disconfirmation by rules concerning exceptions.

5. A set of synchronic and diachronic rules, organized in a default hierarchy, gives rise to an emergent mental model. The mental model guides behavior and serves to generate the predictions that serve as the basis for inductive change.

6. Rules act in accord with a principle of limited parallelism. Those rules with their conditions satisfied by current messages compete to represent the current state of affairs and to guide thinking and action. But in addition to competing with each other, multiple rules will often act simultaneously to complement and support each other. Through summation of converging evidence, the system can use multiple sources of weak support to arrive at a condfident conclusion.

7. Induction involves two basic classes of mechanisms: (a) mechanisms for revising parameters such as the strength of existing rules and (b) mechanisms for generating plausibly useful new rules.

8. Mechanisms for generating new rules are constrained by triggering conditions that tend to ensure that new rules are likely to be useful to the system. Most particularly, inductions are triggered in response to the consequences of the use of current knowledge, such as failed or successful predictions.

9. Induction is guided by background knowledge about the variability of classes of objects and events. It follows that a major goal of inductive systems is to learn about the variability of the environment.

9.3 Comparison with Alternative Approaches

New ideas arise from recombination of the old[3]. We have attempted to give substance to this piece of conventional wisdom by modelling how such recombination takes place,

[3]This section is taken from pp. 22–27 of *Induction*.

when it takes place, what is recombined, and what created. Our framework itself, of course, is the product of inductive mechanisms; our goal, one might say, is a theory that can account for its own origin. Our framework has many ties to earlier theories of cognition. What is it that we believe is novel about the ideas we will be presenting?

The answer lies in the integrative recombination of theoretical concepts through the use of a genetic algorithm. The pieces required to understand induction have lain close at hand, but certain critical relationships among these pieces have been missing. Default hierarchies have been discussed, but typically without reference to induction. Schemas and mental models have also been developed as theoretical concepts, but not as structures that emerge from combinations of rules. Rule-based systems have been in existence for almost two decades, but have generally been used for serial reasoning rather than for exploring the possibilities of parallel competition and collaboration among rules. The variability of the environment customarily has been viewed as an unfortunate pitfall for cognitive systems, rather than as a source of information to guide intelligent adaptation.

In order to locate our framework with respect to alternatives, it may be useful to contrast our approach briefly with two major theoretical points of view that are currently at the leading edge of cognitive theory. These are production systems and connectionist networks.

9.3.1 Contrast with Standard Production Systems

In their performance aspects the systems that fall within our framework have most in common with standard production systems, which are also based on the cyclic matching and execution of condition-action rules (Newell 1973; Anderson 1976, 1983). These production systems, however, allow only one rule to fire each cycle, a restriction that places severe strains on both problem solving and induction. An important part of the classifier system framework is the use of parallelism in the form of firing of multiple rules (compare Holland and Reitman 1978; Holland 1986; Rosenbloom and Newell 1986; Thibadeau, Just, and Carpenter 1982).

The restriction to firing only one rule tends to lead to sets of rules in which it is necessary to have a single rule available to handle any conceivable step in a problem situation. This in turn creates problems in dealing with novel situations that do not match any useful rule. Suppose, for example, the system contains the rule "If the goal is to identify an animal, and it barks and wags its tail, then declare it is a dog"; and suppose that an animal is heard to bark, although its tail is unseen. The observation would fail to match the rule, and unless some alternative rule is available the observation would be left unidentified.

Intuitively, however, the system ought at least to generate a conjecture: the animal *might* be a dog. To capture this intuition, partial matching is sometimes allowed (Anderson 1983): a rule may sometimes fire when only *some* of its conditions are matched.

This appears to produce the intuitive result for the above example. However, allowing rules to fire when only some of their conditions are matched can have treacherous consequences. Suppose the system has the rule "If the goal is to cross a body of water, and it is about a mile wide, and you are a strong swimmer, then decide to swim across". Now imagine that our system, a lamentably weak swimmer, wants to cross a mile-wide lake. A partial match will be found, syntactically identical to that in the "animal" example (two of three conditions in the most relevant rule are satisfied). The result, sad to say, is a dramatic decrease in our system's life expectancy.

The problems created by partial matching can be avoided in a system with greater parallelism. Where there are somewhat useful general rules, such as "If an animal barks, then it is a dog", these can coexist with more valid specific rules, such as "If an animal barks and wags its tail, then it is a dog". Performance will be dominated by the more specific rule (perhaps with support from the more general rule when the specific rule is matched; but the general rule will provide *some* information even when the specific rule does not apply. General rules will only be maintained if they are useful, however. A rule such as "If you want to cross a wide body of water, then swim across", which fails to specify a critical prerequisite for the suggested action (the ability to swim), would be weeded out by inductive mechanisms (preferably using "lookahead" rather than an overt test!). In general, novel situations can be accommodated by maintaining useful but fallible general rules (that is, rules with few conditions), as well as by providing the system with procedures for finding and using analogies, without having to resort to partial matching.

Another criticism that has been directed at standard production systems is that they treat intelligence primarily in terms of quasi-linguistic representations and inference processes akin to those associated with conscious thinking (for example, Newell and Simon 1972; Anderson 1983). Hofstadter (1985, chap. 26) has distinguished between "cognitive" and "subcognitive" approaches to the understanding of intelligence. Subcognitive approaches assume highly parallel processing of small units of information, taking place below the level of conscious awareness. Many theorists have argued that a great deal of mental processing, underlying activities ranging from early visual pattern recognition to more abstract categorization and hypothesis-generation tasks, takes place at the subcognitive level. Standard production systems, because of their seriality and their use of abstract linguistic categories in the conditions of rules, do not appear to offer a description of subcognitive processing. Needed instead are more flexible systems in which categories can emerge through inductive mechanisms without being built in by the programmer.

9.3.2 Contrast with Connectionist Models

A desire to represent subcognitive processes, together with considerations of neurophysiological plausibility, has fueled recent interest in "connectionist" representation schemes.

Connectionist models describe mental processes in terms of activation patterns defined over nodes in a highly interconnected network (Hinton and Anderson 1981; Rumelhart, McClelland, and the PDP Research Group 1986). The nodes themselves are elementary units that do not directly map onto meaningful concepts. Information is conveyed not by particular individual units but by the statistical properties of patterns of activity over collections of units. An individual unit typically will play a role in the representation of multiple pieces of knowledge. The representation of knowledge is thus parallel and distributed over multiple units.

In a connectionist model the role of a unit in mental processing is defined by the *strength* of its connections—both excitatory and inhibitory—to other units. In this sense "the knowledge is in the connections," as connectionist theorists like to put it, rather than in static and monolithic representations of concepts. Learning, viewed within this framework, consists of the revision of connection strengths between units.

As has been the case for production systems, connectionist models are currently proliferating, and it would be premature to strictly delimit the principles embodied in connectionist schemes or to attempt to decide their eventual status relative to production systems. Connectionist models clearly have a very different flavor from prototypical production systems, and in many respects this flavor is quite consistent with the rule-based framework we are proposing. We share with connectionist models such general principles as integration of multiple sources of evidence, revision of strength parameters on the basis of feedback derived from performance, and emergence of higher-order structures from more elementary components. We assume rules can have inhibitory effects akin to the inhibitory connections postulated in network models such as those of McClelland and Rumelhart (1981).

Our framework differs from connectionism, however, in several important respects. Our rule-based system has two fundamental inductive processes at its disposal: revision of the strengths of existing rules and generation of new rules. In contrast, current connectionist models learn exclusively from revision of connection strengths between elementary units. The models exhibit a property that might be termed "topological nativism": it is assumed that the entire network of potential connections between units is part of the cognitive system's neural hardware. Many of these connections may simply be latent, with zero strength; nonetheless, all learning consists of revision of the strengths of existing connnections rather than the building of new ones.

This is not to say that connectionist models do not learn the equivalent of new rules; they can indeed accomplish this task by strength-revision procedures that establish new stable patterns of activation over *sets* of units. The implicit generation of a new "rule" is thus treated as the outcome of revising the strengths of a large set of innate connections. In contrast, our framework treats rule generation more as a "quantum" knowledge change than as incremental strength revision. Basic inductive mechanisms such as generalization are capable of linking previously unconnected elements in a new rule that may immediately produce radical behavioral changes.

9.4 Q-morphisms

A classifier is a rule that attempts to match a message on the classifier system's message list. Messages are bit strings, but classifiers may contain 0's, 1's and #'s, where the "#" matches any character. Given two classifiers C_1 and C_2, where C_2 is like C_1 except that some of the 0's and 1's in C_1 are replaced by #'s in C_2, we can say that C_2 is more general than C_1, since it will match every message that C_1 will match and others as well. C_2 may be regarded as the default response to the stimulus that both match, while C_1 may encode the appropriate response for more specific cases. The mechanism in our classifier systems that determines which matching classifiers will fire at any given moment favors those that are more specific over those that are more general, allowing default rules to fire in general cases but overriding them when exceptional stimuli are presented. In what follows, we show formally that this is a useful technique to adopt.

In practice a model of a complex environment will typically prove to be less than completely valid[4]. Such a situation is depicted in Figure 9-1[5].

In Figure 9-1, P_2 is chosen to distinguish S from other members of the equivalence class induced by P_1. Then T'_2 is chosen to map the image of S under P_2 into the image of $T[S]$ under P_2. Under this arrangement the elements of level 1 serve as defaults, to be used unless the exception is detected, at which point level 2 is invoked. For example: The level 2 element might be "small, striped, fast-moving, airborne" (*e.g.*, a "wasp") that remains "fast-moving" (*i.e.*, it continues flying), thus contradicting rule T'_1, which predicts a slowdown. T'_2 compensates for this exception by predicting no slowdown in the specific case of "small, striped, fast-moving, airborne" objects.

Here one element of the category of "fast-moving objects" violates the expectation that fast-moving objects will invariably slow down: some fast-moving objects ("wasps") may not slow down in most circumstances. The level 1 model makes erroneous predictions with respect to these objects because it does not distinguish them from other members of the equivalence class.

Level 2 of the model, depicted in Figure 9-1, corrects for these exceptions. We can think of the mapping P_2 as based upon additional properties that enable a distinction to be made between wasps and other objects. For example, an additional detector d_4 might encode the property "airborne", allowing P_2 to define the more specific category of "small, striped, fast-moving, airborne objects". Once this new category is available in the elaborated model, the model transition function can also be refined by the addition of T'_2, which indicates that small, striped, fast-moving, airborne objects continue to move quickly, rather than slowing down.

The model depicted in Figure 9-1 is a layered set of transition functions, which we will term a *quasi-homomorphism*, or simply *q-morphism*. In a q-morphism a higher layer in the model, with its broader categories, provides default expectations (such as,

[4]This section is from p. 34 of *Induction*.

[5]Figure 9-1 reprinted by permission of MIT Press.

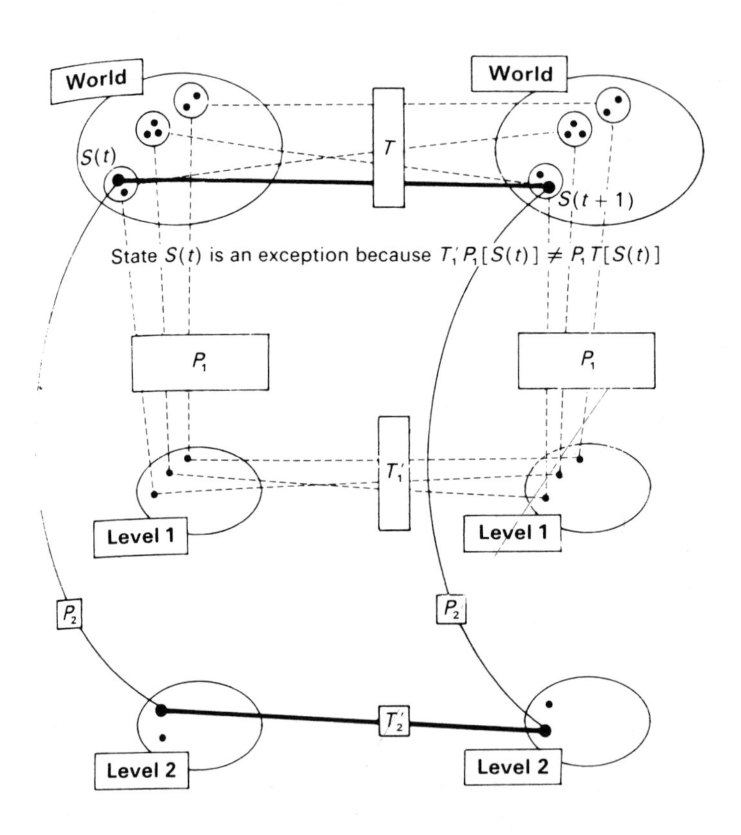

Figure 9-1: Q-morphism (defaults only).

fast-moving objects slow down) that will be used to make predictions unless some exceptional category is signaled. An exception evokes a lower level of the model, at which a different model transition function is specified to capture the exception. The relations between level 1 and level 2 of the model in Figure 9-1 can be generalized to more complex multilayered q-morphisms (see proof below). For example, the level 2 model will itself admit of exceptions (striped marbles thrown through the air slow down.)

9.5 Q-Morphisms and Induction

The process of model construction can be viewed as the progressive refinement of a q-morphism[6]. The initial layer of the model will divide the world into broad categories that allow approximate predictions with many exceptions. Each additional layer in the hierarchy will accommodate additional exceptions while preserving the more global regularities as default expectations. The induction process will be guided by failures of the current model, in that failed expectations will serve as triggering conditions for the generation of new, more specialized rules. In a complex environment the process of model refinement by the cognitive system is unlikely ever to be completed. New exceptions to the current model will always be possible.

The concept of a q-morphism captures several basic aspects of a pragmatic account of the performance of cognitive systems. First, its hierarchical structure allows the system to make approximate predictions on the basis of incomplete knowledge of the environment. Second, as the model is refined, rules that represent useful probabilistic regularities can be retained as defaults. The following proof demonstrates that a hierarchy of default rules with exceptions can represent knowledge more concisely (that is, with fewer total rules) than a system restricted to "exceptionless" rules.

The representation of events and event sequences at different levels of a default hierarchy is the most basic way for a system to deal with variability within a class of events. Events are treated as equivalent unless there are features — or a failed expectation — suggesting further differentiation.

9.6 Generalized Definition of Q-Morphisms

The full variety of q-morphisms can be defined precisely as follows[7]. The q-morphism can have any finite number of layers r. For each layer j, $1 < j < r$, there is a map P_j that aggregates the set of world states W into the model states M_j assigned to that level, $P_j : W \rightarrow M_j$. There will be one new state in the model for each equivalence class in W induced by P_j. P_j is usually a partial map defined only for some elements of W (the "exceptions")—that is, it only selects certain subsets of W. The construction of the

[6]From pp. 34–36 of *Induction.*

[7]From p. 65 of *Induction.*

model is cumulative, so we define a function P_j* that subsumes the total structure up to level j. We do this for any level j by using the cumulative map from the previous level, $P_{j-1}*$, along with the "local" map for the current level, P_j. Letting w be an arbitrary world state, and letting m and m' be distinct model states from M_j at level j, we proceed as follows:

1. if $P_{j-1} * (w) = m'$ and $P_j(w) = m$, then $P_j * (w) = P_j(w) = m$; that is, the new "exception" overrides the old "default".

2. Otherwise, $P_j * (w) = P_{j-1} * (w)$; that is, when P_j does *not* supply a new image, the old one is used.

We start the process by setting $P_1* = P_1$. For simplicity and completeness we require P_1 to map *every* element of W into some element of M_1 — that is, the equivalence classes inducted by P_1 partition W.

We define the total set of states of the q-morphism up to level j, Q_j, by adding the model states from the current level, M_j, to the cumulative set of states from the previous level, $Q_j = M_j \bigcup Q_{j-1}$. For certain states of the world the transition function T'_j will make a valid prediction; that is, for such states T'_j will satisfy the conditions for a homomorphism,

$$T'_j P_j * (w) = P_j * T(w).$$

Define W_j* to be the set of world states for which the q-morphism makes correct predictions at level j. Then the layered set of model transition functions $\{T'_1, T'_j, T'_r\}$ is a *q-morphism* if, for every level j, $1 < j < r$, the set of states W_j* yielding correct predictions *properly* includes the corresponding set of states $W_{j-1}*$ from the preceding level. In short, each layer added to the model adds to its ability to make correct predictions. The successive layers of the q-morphism constitute a *default hierarchy* in which Q_{j-1} is the *default set* for Q_j and Q_j is the *exception set* for Q_{j-1}.

9.7 The Economy of Defaults and Exceptions: The Parsimony of Q-Morphisms

Consider, first, a homomorphic model of the environment based upon a set of k binary detectors[8]. The model is specified by two maps:

1. a mapping P from the world states W to a set M of 2^k model states, one model state for each k-tuple of detector readings;

2. a transition function T'_M providing commutativity of the diagram.

[8]From pp. 66-67 of *Induction*.

For the purposes of this argument, we will assume that there is no simpler homomorphic model using some subset of the k binary detectors; that is, any model using fewer of the binary detectors fails to meet the "commutativity of the diagram" requirement and hence produces errors. A straightforward rendering of T'_M in terms of rules then requires one condition-action pair for each state of the model. That is, for each s in M there is a rule $s \rightarrow s'$ that provides the transition $T'_M(s) = s'$. Because there are 2^k states, 2^k rules are required to implement T'_M.

Consider, next, a nontrivial q-morphism model based on the same set of detectors. Let the level 1 map P_1 be based on $k_1 < k$ detectors. Consider a particular rule C in the representation of the transition function T'_1. The condition for C actually responds to 2^{k-k_1} distinct k-tuples of detector values. That is, the rule responds to 2^{k-k_1} distinct states of the detailed model (P_1 induces equivalence classes in the detailed homomorphic model). Because we are dealing with a q-morphism, C will provide the correct transition, the one specified by the homomorphic transition function T'_M, for only a proportion of $d < 1$ of the 2^{k-k_1} states satisfying its condition. These errors must be corrected by "exceptions" specified by deeper levels of the q-morphism. In the worst case each of the errors will have to be corrected by rules involving all k detectors. The rule C together with the $(1-d)(2^{k-k_1})$ k-detector exceptions then yields the same transitions as the set of 2^{k-k_1} rules from the homomorphic mode. In other words, $(1-d)(2^{k-k_1}) + 1$ rules of the q-morphism replace 2^{k-k_1} rules of the homomorphic model, a saving of $d(2k - k_1) - 1$ rules.

As an example, let the homomorphic model use 10 detectors ($k = 10$), let layer 1 of the q-morphism be based on the values of 5 detectors, ($k_1 = 5$), and let the rule C be correct only half the time ($d = 0.5$). Then, at worst, the q-morphism requires 17 rules to accomplish the same transitions as 32 rules from the homomorphic model, a saving of 15 rules. If all rules in the q-morphism have error rates no higher than C, then the q-morphism model would use at most 17/32 as many rules as the homomorphic model.

Because the number of rules required to "cover" a given set of states increases exponentially with the number of detector values specified in their condition parts, the savings of a q-morphism increase dramatically as the number of layers increases. For example, consider a q-morphism in which the highest layer uses 2^{k_1} rules based on k_1 detectors, and each successive layer uses rules employing k_1 additional detector values to correct for exceptions in the previous layer. Let rules at each level, except the deepest level, be correct over at least a proportion d of the instances satisfying them. (We assume the rules at the deepest level deal correctly with individual states of a homomorphic image, as in the earlier example.) Let the q-morphism involve n layers so that the deepest level involves nk_1 detectors. Then, following an argument that parallels the one above, we see that the q-morphism requires at most $\sum_j (1-d)^{j-1} 2^{jk_1}$ rules to specify the same transition function as would be specified by 2^{nk_1} rules in the detailed model. For a 5-layer model ($n = 5$) with 2 additional detectors being used in each successive layer ($k_1 = 2$), and an error rate of 0.5 ($d = 0.5$), the q-morphism requires less than 124 rules whereas the direct presentation of the homomorphic model would require 1,024 rules. For a still larger model, $n = 10$, under the same conditions, $k = 2$, $d = 0.5$, the q-morphism requires 4,092 rules compared to over 1,000,000 rules required for direct presentation.

The q-morphism offers still greater parsimony if the corresponding model simply ignores some of the rarer exceptions, permitting a (low) error rate under such circumstances. Since models of real situations will almost always be prone to error, no matter how refined, the model-building process can concentrate on building exception rules for costly errors only, with still further savings in the number of rules required.

9.8 Genetic Operators in Classifier Systems

New rules can be generated by recombination only if the condition and action parts of rules can be decomposed into simple components that can be selected and combined easily[9]. Hence the system must have a set of building blocks from which any rule can be constructed. The process of generating new rules then pivots on the use of experience to discover and exploit good building blocks. Candidate rules are generated by recombining good building blocks occurring in existing rules that are relevant to the triggering context. Classifier systems, by virtue of their simple representational scheme, allow the exploitation of powerful and general mechanisms for selecting good building blocks for new rules.

How is a cognitive system to rate the components from which its rules are constructed? The strength assigned to a rule provides an estimate of *its* usefulness to the system, but this does not give any direct estimate of the value of its components. There is, however, a simple way, and one quite effective for systems based on competition, to use this information to rate components: the rating of any component is taken to be the average of the strengths of the rules employing it. That is, a good building block is one that occurs in good rules. This is, of course, a crude estimate and one fraught with error in some cases. Nevertheless, the components so highlighted are plausible candidates for use in the construction of new rules, particularly if the system uses the estimates only to bias construction toward the use of such components.

At any given time t there will be several rules in the system employing any simple building block b. That is, the cognitive system has several *instances* of b. As suggested in the previous paragraph, we can assign value $v(b,t)$ to b at time t by averaging the strengths of its instances. For example, let the system contain rules $C1$ and $C2$ with strengths $S(C1,t) = 4$ and $S(C2,t) = 2$, respectively. If these are the only instances of b at time t, then we assign to the building block the value

$$v(b,t) = [S(C1,t) + S(C2,t)]/2 = 3,$$

the average of the strengths of the two instances. The general formula, when b occurs in

[9]pp. 118–121 of *Induction*.

each of $\{C_1, C_2, \ldots, C_n\}$, is

$$v(b,t) = \sum_{j=1}^{n} S(C_j, t)/n.$$

In classifier systems one can designate an important class of building blocks by simply specifying that certain letters from the alphabet $\{1, 0, \#\}$ must occur at certain positions in the condition and action parts of the strings defining the classifiers. We will call such building blocks *substring schemas.* Holland (1975) simply used the term "schema"; we use "substring schema" here to avoid confusion with the schema notion familiar in psychology. Under this interpretation a single rule can be an instance of an enormous number of building blocks: for a classifier condition of length k, the number of potential building blocks is 3^k. Accordingly, it is computationally infeasible to calculate and use the large set of averages $\{v(b,t)\}$ in reasonable amounts of time. The solution is to introduce recombination techniques, termed *genetic operators*, that accomplish this task implicitly by means of selective recombination of parts of rules. The most important genetic operator is *crossover*, which involves the recombination of two rules. Parts of each rule are combined to form a "hybrid" rule that is added to the system to compete with the other rules already there, including the "parent" rules. The offspring displace weak classifiers in the system. If the parent rules are of high strength, then the hybrid offspring is likely to inherit good building blocks from both parents.

Counterparts of this process of emphasis and recombination can be found at any level of abstraction, from the formation of neural cell assemblies (Hebb 1949) to the generation of complex concepts. It is vital to the understanding of genetic algorithms to know that even the simplest versions act much more subtly than "random search with preservation of the best", contrary to a common misreading of genetics as a process primarily driven by mutation. Though genetic algorithms act subtly, the basic execution cycle is quite simple:

1. From the set of classifiers, select pairs according to strength — the stronger the classifier, the more likely its selection.

2. Apply genetic operators to the pairs, creating "offspring" classifiers. Chief among the genetic operators is crossover, which simply exchanges a randomly selected segment between the pairs. Note that crossover can be applied either to *both* the conditions and the actions of the parent rules (as in the generation of couplings, described below) or to only the conditions or only the actions.

3. Replace the weakest classifiers with the offspring.

This simple procedure can be proved to rate and exploit large numbers of building blocks using, implicitly, the averages $v(b,t)$. Surprisingly, more than M^3 building blocks will be usefully processed for every M classifiers processed by the algorithm, a phenomenon known as *implicit parallelism.*

It is instructive to see how crossover can be used to produce couplings. Consider first a simple, direct procedure for coupling a pair of classifiers $\{C_1/M_1, C_2/M_2\}$: Simply replace condition C_2 in C_2/M_2 by the message specification M_1 from C_1/M_1, yielding a new classifier M_1/M_2. (Note that the #'s in M_1 are simply reinterpreted as "don't cares", yielding a valid condition from $\{1, 0, \#\}^k$.) Under this arrangement the message specified by M_2 will be emitted if either C_2/M_2 or M_1/M_2 is activated. But M_1/M_2 will be activated whenever C_1/M_1 is activated, providing the same effect as the direct activation of C_2/M_2 by C_1/M_1.

Now complicate this procedure somewhat by first crossing M_1 with C_2. The "hybrids" H_1 and H_2 produced by the exchange of segments can be treated either as conditions or as message specifications, both being elements of $\{1, 0, \#\}^k$. Adding the pair of classifiers $\{C_1/H_1, H_1/M_2\}$ or the pair $\{C_1/H_2, H_2/M_2\}$ to the system will produce the same coupling effect as the more direct procedure.

What advantage is there to adding the extra complication of crossover to the coupling procedure? The main advantage lies in the fact that the coupling is accomplished via a message specification distinct from M_1 and a condition distinct from C_2, yet with parts from both. This distinguishes activations produced by the new coupling rules from those produced by extant rules. There are serendipitous effects arising from the fact that the new action may "tap into" other conditions that share parts with it, and the new condition may respond to messages from other parts of the system because of shared parts. This would be intolerable if the new rules were other than plausible candidates or if the system were not graceful. As it is, crossover provides the system with a range of associations based on shared components (reminiscent of the role of overgeneralization in the formation of default hierarchies). Moreover, should the coupling prove to be important, the new rules substantially bias future possibilities under the genetic algorithm toward the patterns instantiated by the hybrids. The implicit parallelism of genetic algorithms assures a relatively rapid search of the patterns and combinations involved, leading to the generation of families of tags that produce useful couplings.

George Robertson

Chapter 10
Parallel Implementation of Genetic Algorithms in a Classifier System

10.1 Introduction

Genetic Algorithms are the primary mechanism in Classifier Systems (Holland *et al*, 1986) for driving the selective evolution of rules (learning) to perform some specified task. The Classifier System approach to machine learning, in particular, and Genetic Algorithms, in general, are inherently parallel. Implementations of Classifier Systems and Genetic Algorithms to date have mostly been done on conventional serial computers. Because of this serial bottleneck, researchers have been able to study only relatively small problems. The advent of commercially available massively parallel computers, such as the Connection Machine system (Hillis 1985), now makes parallel implementations of these inherently parallel algorithms possible, and makes it possible to begin studying these algorithms on larger task domains. This paper describes an implementation of Genetic Algorithms in a Classifier System on the Connection Machine.

We begin by describing Classifier Systems as one approach to machine learning, with comparisons to symbolic rule-based systems and Connectionist Network systems. In section 3, we describe a programming style for fine-grained massively parallel computers, called data parallelism (Hillis and Steele 1986). The Connection Machine system is a general purpose data parallel computer, and is described here. In section 4, we describe *CFS, the parallel Classifier System architecture implemented on the Connection Machine system. *CFS is a parallel implementation of CFS-C (Riolo 1986a), a domain independent Classifier System architecture developed at the University of Michigan. In section 5, we describe the implementation of parallel Genetic Algorithms in *CFS, including the parallel data structures and key parallel algorithms used. We summarize by describing the principal results of this work and our plans for further evaluation and research.

10.2 Classifiers Systems and Machine Learning

Classifier Systems represent one basic approach to learning by example. Two other approaches are symbolic rule-based systems and Connectionist Network systems. These approaches, and the inherent parallelism in them, are compared and contrasted with

Classifier Systems below. This provides a framework for machine learning in which to place Classifier Systems, and suggests the ease of implementation for each approach on massively parallel computers.

Symbolic rule-based systems attempt to model high (symbolic) level cognitive processes. A good example is the SOAR problem-solving architecture being studied at Carnegie-Mellon University (Newell), Stanford University (Rosenbloom), and Xerox PARC (see Laird 1985, Laird *et al* 1985, and Laird *et al* 1986). SOAR is an architecture for problem solving and learning, based on heuristic search in problem spaces and chunking. It is based on a modified version of the OPS5 Production System architecture (Forgy 1981). Although SOAR shows promise as an approach to learning, it does not appear to be a good candidate for massively parallel computers. Gupta (Gupta 1984) has done studies of parallelism possible in symbolic production systems and found that systems with more than thirty or so processors would not be effectively utilized. Most of the traditional AI work in machine learning (see Michalski *et al*, 1986) also operates at the symbolic level and is perhaps even less easily parallelized than SOAR.

Connectionist Network approaches to learning began with the work on Perceptrons by Rosenblatt (Rosenblatt 1962) (also see Minsky and Papert 1969). New ways of dealing with the problems that they encountered have recently been studied, including work on Boltzmann Machines (Ackley, Hinton, and Sejnowsky 1985), Back Propagation networks (Rumelhart, Hinton, and Williams 1985, Sejnowski 1986, and Sejnowski, Keinker, and Hinton (in press)), work by Klopf (Klopf 1982), Barto (Barto 1985), Grossberg (Grossberg 1978), Feldman (Feldman 1982), Hinton (Hinton 1981), and Minsky (1986). These systems model low level neural processes, with long-term knowledge represented as strengths (or weights) of the connections between simple neuron-like processing elements. The Boltzmann Machines introduce a stochastic process to avoid the pitfalls of local minima in learning behavior. Back Propagation networks have successfully been used to learn to recognize symmetries in visual patterns (Sejnowski, Keinker, and Hinton in press), as well as to learn a speech synthesis task (Sejnowski 1986). In this latter task a corpus of text and its phonetic translation is used as the example for learning. A fixed initial structure (a three level Connectionist Network) begins with random link and node threshold weights and over time learns the proper weights to translate the text into understandable phoneme streams. These systems are well-adapted to massively parallel systems; in fact, Back Propagation and Boltzmann Machines have been implemented on the Connection Machine system.

Classifier Systems attempt to model macroscopic evolutionary processes with Genetic Algorithms and natural selection. Holland's early work on Genetic Algorithms (Holland 1975) has recently led to numerous research efforts and application of Classifier Systems to a number of task domains (see Forrest 1985, Holland 1985, Holland 1986, Riolo 1986a, Riolo 1986b, Wilson 1986, and Wilson 1985.) This approach shares some properties with both the symbolic rule-based and Connectionist Network approaches. It is rule-based, but the rules are low level message passing rules (below the symbol level). A symbolic level can be added on top of Classifier Systems (see Forrest 1985). Like Boltzmann Machines, Classifier Systems use a stochastic process to avoid local minima in learning behavior. They use Genetic Algorithms to replace weak rules (those not contributing or contributing incorrectly to the problem solution) with offspring of strong rules. Unlike the Back Propagation approach, which requires predefined layers of networks, the Classifier System approach does not require any predefined structure (although one can be provided if desired, in the form of an initial set of rules). Because of the nature of the rules, messages, and the match cycle, this approach is very well-adapted to massively parallel computers.

To date, work on all of these approaches has taken place primarily on conventional serial computers with simulators. Because the speed of these serial simulators depends directly on the size of the problems being solved, only small examples have been used to test these ideas. The massive parallelism and dynamic reconfigurability of the Connection Machine offer a chance to implement and evaluate the Connectionist Network and Classifier System approaches to learning on much larger (and more realistic) task domains. The speed of the parallel versions of these systems is nearly independent of the size of the problems being solved.

10.3 Parallelism on the Connection Machine

Most computer programs consist of a control sequence (the instructions) and a collection of data elements. Large programs have tens of thousands of instructions operating on tens of thousands, or even millions of data elements. There are opportunities for parallelism in both the control sequence and in the collection of data elements. In the control sequence, it is possible to identify threads of control that could operate independently, and thus on different processors. This approach is known as "**control parallelism**", and is the method used for programming most multiprocessor computers. The primary problems with this approach are the difficulty of identifying and synchronizing these independent threads of control. Alternatively, it is possible to take advantage of the large number of data elements that are independent, and assign processors to data elements. This approach is known as "**data parallelism**" (Hillis and Steele 1986). This approach works best for large amounts of data, and for many applications is a more natural programming approach. The Connection Machine system is a general purpose implementation of data parallelism.

The Connection Machine system is a dynamically reconfigurable computer with 65,536 processors and 32 Mbytes of memory, which can process large volumes of data at speeds exceeding one billion instructions per second (1000 MIPS). This computer is a departure from the conventional von Neumann model of computing, which has one processor with a large memory. In the Connection Machine, the processing power is distributed over the memory, so that there are many processors, each with a small amount of memory (4096 bits).

In addition to a large number of processors, a data parallel computer must have an effective means of communicating between processors. The Connection Machine system has a communications system that allows any processor to send a message to any other processor, with possibly all processors sending messages at the same time. This mechanism allows applications to dynamically reconfigure the communications topology to adapt it to the communications needs of the moment. For example, in the low level part of an image understanding system, a two dimensional grid is the most appropriate topology. At the intermediate level, such a system might want to communicate in a tree structured topology. And at the highest level, a semantic network for understanding might require an arbitrary network topology. Each of these communications topologies can be dynamically configured on the Connection Machine.

From a programming point of view, the Connection Machine system can be thought of as a parallel processing accelerator for a conventional serial computer. In fact, the Connection Machine system is made up of two parts: a front-end computer (like a Vax or a Symbolics 3600), and an array of Connection Machine processors and memory. The front-end computer can operate on the Connection Machine memory as though it were part of its own memory, or it can invoke parallel arithmetic, logical, or communications (data movement) operations on that memory. The program runs on the front-end computer, invoking parallel operations when necessary. Thus, all the program development tools of the front-end computer are available for developing Connection Machine programs.

Connection Machine programming languages are simple parallel extensions to familiar sequential languages. C* (Rose and Steele 1986) is a parallel version of the C language (Harbison and Steele 1984), and *Lisp (Lasser 1986) is a parallel version of Common Lisp (Steele 1984). In each of these languages, the extensions were made in an unobtrusive manner. For example, in *Lisp, two parallel variables can be added together with +!! (the !! at the end of a operator name indicates the parallel version of that operator). The process of writing a data parallel program with a language like *Lisp is quite simple: write an algorithm as though it were operating on one data element, then use the parallel versions of the operators. In control parallelism, understanding how to do a parallel task decomposition is often an intellectually difficult task. In such a system, it is also often difficult to take an existing program and understand it. However, in data parallelism, the programming style is much closer to what programmers are used to, and it is thus much simpler to produce and understand data parallel programs.

10.4 *CFS: A Parallel Classifier System

The state of the art for Classifier Systems is represented by the CFS-C system (Riolo 1986), designed and implemented at the University of Michigan. This system is designed with parallelism in mind, but has been simulated on conventional serial machines, and thus has been restricted to small task domains. Serial implementations of CFS-C are relatively fast for small sets of classifiers; up to 20 cycles per second for 200 Classifiers on a one MIP (one million instructions per second) computer. However, because the match procedure (described below) requires every message to be matched against every classifier, serial implementations for large sets of classifiers are too slow to be useful.

*CFS is an implementation of CFS-C on the Connection Machine, written in *Lisp. Because of the parallel nature of *CFS on the Connection Machine, it will work with 65,000 classifiers about as fast as it works with 200 classifiers. The speed of the system does not depend on the number of classifiers and is relatively fast; up to 10 cycles per second for 65,000 Classifiers. Thus, *CFS on the Connection Machine provides a way to explore and evaluate Classifier Systems and Genetic Algorithms on large task domains.

10.4.1 *CFS: Overview of Operation

From the user's point of view, both CFS-C and *CFS provide the same basic structure; a message-passing rule-based system that uses Genetic Algorithms to evolve rules to solve some specified problem. Both systems are task domain independent. To define a task domain, you simply define three functions: a function to provide input messages that describe the state of the external environment on each cycle (these are called "Detector" messages); a function to analyze output messages to alter the external environment (called "Effectors"); and a function to evaluate the changes made to the environment, and supply a reward or punishment for those changes.

A "Message" is a fixed length bit string, with each bit position containing a zero, one, or "don't care" value. A "Classifier" is a rule with two conditions and an action. The conditions are message patterns which are matched with incoming messages. The action is a message pattern for production of an outgoing message.

On each cycle, the messages output from the previous cycle are combined with Detector messages describing the environment to provide the incoming message list. All Classifiers are matched with all the messages on the message list. The matching Classifiers can each post one or more new messages to the outgoing message list. The message list is limited to a small size to force the Classifiers to compete for the right to post new messages. There is a strength associated with each Classifier that is used to control this bidding process. The strength reflects how good a rule is; that is, a high strength Classifier is one that contributes to a correct solution, while a low strength Classifier is one that either does not contribute or contributes to an incorrect solution. Classifier strength is adjusted with several different mechanisms: (1) reward or punishment from the evaluator changes the strengths of all Classifiers that won the bidding competition and posted messages during that cycle; (2) Classifiers that win the bidding process pay their bid to the Classifiers in the previous cycle that produced the messages which the winning Classifiers matched (this is called the Bucket Brigade algorithm (Holland 1985), and is necessary for chains of rules to form and survive); and (3) taxes are used to eliminate non-contributors and to help prevent over-general Classifiers from dominating the bidding process.

Genetic Algorithms are used periodically to replace some percentage of the population of Classifiers (generally, low strength Classifiers are replaced) with offspring from matings of other (generally, high strength) Classifiers. The two primary Genetic Algorithms used are Crossover Mating and Mutation. The Crossover algorithm considers the entire Classifier (both conditions and the action) as a chromosome, picks a random point for the crossover, and crosses two parents to produce two offspring (i.e., the high order bits up to the randomly picked point are swapped in the parents). The Mutation algorithm uses a Poisson distribution to decide whether to make zero, one, two, or three random modifications to a new offspring. The mutation rate is kept quite low, so that only a few offspring are mutated at all, and very few have more than one mutation.

10.4.2 *CFS Display System

The display system of *CFS is part of a Learning Laboratory being developed to study different approaches to machine learning on the same or similar task domains. The Learning Laboratory has two main features: (1) a learning approach-independent way of describing task domains; and (2) a user interface that allows dynamic control over numerous graphic displays, to aid in the debugging and understanding of the behavior of various learning systems. Both *CFS and CFS-C have large parameter spaces to control learning performance. There are no good guidelines for how to tune these parameters and little is known about what kinds of tools are needed to guide tuning them. The Learning Laboratory display system provides a beginning of a solution for dynamic debugging and tuning of such systems.

Figure 10-1 shows the output of one learning experiment with *CFS. The task is letter sequence prediction (Riolo 1986); in this case, the sequence "a b c d" is repeated over and over, and the system must learn to predict which letter comes next given a description of the current letter and up to three of the previous letters in the sequence. The Learning Laboratory display system provides from one to six display panes, and allows the user to dynamically change display pane configurations and the type of graph displayed in each pane. For this figure, there are four display panes. The first contains the learning curve, which shows the percentage of correct guesses over time. Time in this experiment is measured as the number of cycles, with one cycle for each letter in the sequence. As you can see, for this example the system learned to predict the sequence in about 500 cycles. The second display pane contains a speed curve, which shows the speed of the system in cycles per second over time. The third display pane contains the average Classifier strength over time. The drop in strength at the beginning reflects the effect of the taxes weeding out non-contributing Classifiers. The fourth display pane shows a histogram of Classifier strength. Finally, at the bottom is a user interaction pane (Lisp Listener), where a summary of the learning performance of this experiment

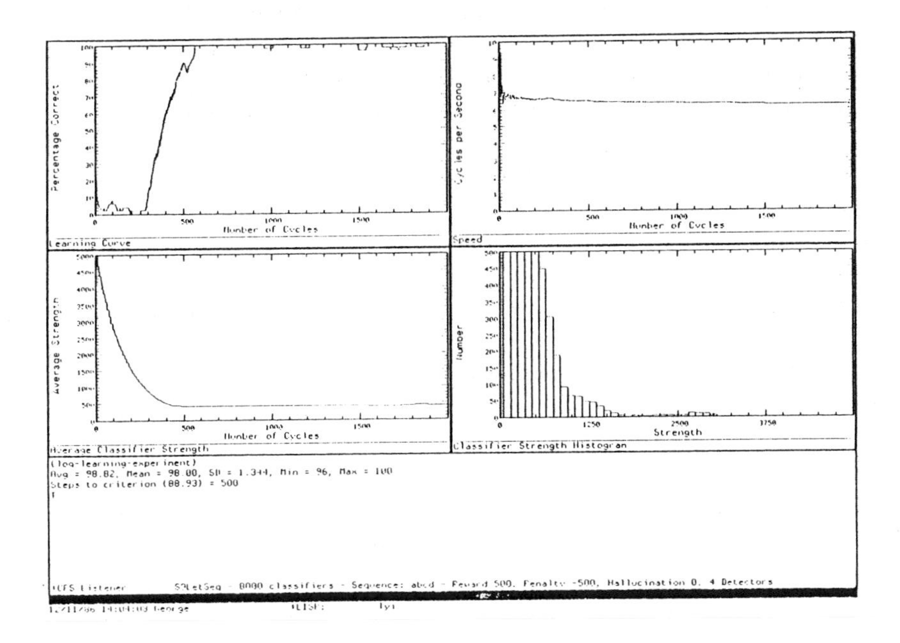

Figure 10-1: Letter Sequence Prediction with *CFS

is displayed. This summary information is recorded in log files so that it can be used to compare the effects of different parameter settings, different Genetic Algorithms, and different approaches to machine learning. These types of display help enormously in understanding and tuning the behavior of systems like *CFS.

10.4.3 *CFS: Parallel Data Structures and Algorithms

The inherent computational load of *CFS can be characterized as proportional to the product of the length of the message list and the number of Classifiers. Experience suggests that the message list should be restricted to a small size to promote strong competition in the Classifier population. It also appears that solving problems in large real-world task domains will require a large Classifier population. These two observations, along with the fact that the Classifiers are operated on almost entirely independently, suggest a natural data parallel approach for *CFS, which is the assignment of one processor to each Classifier.

The primary parallel data structures in *CFS are associated with Classifiers. In addition to the two conditions, action, and strength that have already been mentioned, there are about 30 other variables associated with each Classifier. Some of these variables maintain performance statistics, others are used to control various algorithms in the system. Values stored in parallel variables on the Connection Machine can be of any size needed. The parallel variables that represent a Classifier include several one bit booleans, signed integers ranging from 5 to 32 bits, and unsigned integers that are as long as the message size in the particular task domain (9 bits for the letter sequence prediction task). In the current implementation, about 850 bits are used for each Classifier, with performance statistics accounting for about half of that. That means that several Classifiers could be implemented on each physical processor on the Connection Machine (since each has 4096 bits of memory). In fact, the Connection Machine has a Virtual Processor mechanism that supports that. By selecting the right virtual processor configuration, a 65,536 processor Connection Machine can simulate a million processors, hence a million Classifiers.

The other primary data structure in *CFS is the message list, which is maintained on the front-end computer. Although all operations on Classifiers are done in parallel, the operations on messages in the message list must be done serially. However, the size of the message list is relatively small (less than 30 messages for most applications), and the number of operations that sequence through the message list is small (the match procedure, the creation of the new message list, and activation of Effectors).

To see how the parallel Classifiers and the serial message list interact, let us consider the match algorithm. Assume that each message is N bits long, and that there are at most M messages in the message list. The conditions in the Classifiers are represented as two N-bit unsigned fields, one for the bits (zeros and ones) and the other for the wildcards (the "don't care" bits). Let us allocate an M-bit "message winner" field for each condition, to represent which messages matched that condition (i.e., the m'th bit of that field will be one if the m'th message in the message list matched that condition). To match all the messages with all the Classifiers, we first sequence through the message list broadcasting each message to all Classifiers in parallel. This is the only serial part of the match, the rest of the operations are done in parallel. We compare the logical OR of the wildcard bits and the condition bits with the logical OR of the wildcard bits and the message bits. If they are the same, we set the m'th message winner bit. After sequencing through the message list, we find all Classifiers that matched by selecting all Classifiers that have non-zero message winner fields for both conditions. The whole match process takes about 3.5 milliseconds plus about 0.5 milliseconds for each message on the message list, regardless of the number of Classifiers (up to 65,536).

10.5 Parallel Genetic Algorithms In *CFS

As mentioned earlier, the critical component of Classifier Systems that drives their learning behavior is the Genetic Algorithms. In *CFS, there are currently two parallel Genetic Algorithms employed: Crossover Mating and Mutation. These algorithms are invoked periodically and replace some percentage of the population of Classifiers with offspring of matings of other Classifiers. In the experiment shown in Figure 10-1, the Genetic Algorithms were invoked once every eleven cycles, with five percent of the population being replaced each time. The frequency of invocation of Genetic Algorithms is a prime number to avoid the effects of cyclic patterns in the test environment, which might cause strength patterns that alter the effectiveness of the Genetic Algorithms. The percent of the population being replaced is relatively low so that the Bucket Brigade and other strength adjustment algorithms have time to establish some stability.

The first step in these Genetic Algorithms is to pick a set of Classifiers to be replaced and a set of Classifiers to be parents. These two sets are the same size and a one-to-one mapping is established between them, in the form of two-way pointers between elements of each set. The parents are then copied onto the Classifiers being replaced. From that point on, the set of Classifiers being replaced is referred to as the set of offspring. Some percentage of the offspring are left as pure replications, while the rest are paired and mated with the crossover algorithm. Finally, there is a small probability that each new offspring will be changed slightly by the mutation algorithm. All of the algorithms described in this section operate at speeds that are independent of the number of Classifiers involved (up to 65,536).

10.5.1 Picking Parents and Classifiers to Replace

Relatively low strength Classifiers are picked to be replaced. The choice is made probabilistically, making random draws without replacement from the population with the probability of being picked proportional to the square of the inverse of the strength of the Classifier. This is done so that weak Classifiers have some small chance of not being replaced, and strong Classifiers have some small chance of being replaced. Likewise, relatively high strength Classifiers are picked as parents. This is also done probabilistically, with the probability proportional to the square of the strength.

A "**Parallel Random Weighted Selection**" algorithm supports the picking of parents and Classifiers to replace. Of the several approaches to implementation of this algorithm, the following has been the most successful. First, generate a random number between zero and the weight (*e.g.*, the square of the strength) for each Classifier in parallel. Then, **Rank** those random numbers in parallel. The **Rank** operation is the first step in a parallel sort, and is done in log time on the Connection Machine using either Batcher's bitonic sort (Batcher 1968) or a radix sort (Hillis and Steele 1986). After ranking the random numbers, the Classifier with the smallest random number will have rank 0, the Classifier with the largest random number will have rank N (the size of the population), and each rank will be unique. At this point, select the lowest five percent (or whatever percent is being replaced) of the ranked Classifiers as the Classifiers to replace, and the highest five percent of the ranked Classifiers as parents. This random weighted selection algorithm takes about 23 milliseconds on the Connection Machine.

10.5.2 Crossover Mating

At this point, we have identified two sets of Classifiers: a set of parents and a replacement set. Before proceeding, we need to replace the Classifiers picked for replacement with copies of their parents. The first step in this process is to make each offspring (one of the replacement set) point to a parent, and vice-versa. This is done using a "**Parallel Rendezvous**" algorithm, which takes three steps:

1. **Enumerate** the parent set, and **Send** the processor address of each parent to a rendezvous variable in the enumerated processors. Enumeration is another log time operation on the Connection Machine. The result of enumerating the parent set will be a unique numbering of the parents from zero to the number of parents. The **Send** from the parents to the enumerated processors will result in the rendezvous variable in processor zero containing the processor address of the first parent, in processor one the rendezvous variable will contain the processor address of the second parent, and so on. If there are N parents, then the rendezvous variable in the first N processors will contain the addresses of the parents.

2. **Enumerate** the replacement set, and have each offspring **Get** the address in the rendezvous variable from the enumerated processors. Since this enumeration will be the same as the replacement rank developed while picking the Classifiers to replace, we can use the replacement rank and skip this enumeration to save time. The **Get** will result in the first offspring obtaining the address of the first parent from the rendezvous variable in processor zero, the second offspring obtaining the address of the second parent, and so on. Now, each offspring has a pointer to its parent.

3. Offspring **Send** their processor addresses to their parents. Since offspring now have their parents' addresses, they can do this directly.

Given the pointers between parents and offspring, we now replace the offspring with copies of their parents. This is done by having parents **Send** relevant parts of themselves to their offspring using the offspring address pointers derived from the rendezvous step. Some parallel variables associated with the offspring are simply initialized, while others are copied from their parents. The strength of a new offspring is set to a value half way between its parent's strength and the average strength of the population. This allows the new offspring to participate in the bidding process quickly, without dominating the process.

The Crossover algorithm used in *CFS allows some percentage (a system parameter) of null crossovers, or pure replications. To decide which offspring will be replications, we generate a random number between 0 and 100 for each offspring. If that number is below the specified cutoff, we are done with the Crossover algorithm, since the parent has already been copied.

For the remaining offspring, we proceed with the Crossover algorithm by picking pairs to mate and crossing those mates at random points. The next step is to pair the offspring (since these are now copies of parents, this is equivalent to pairing the parents). This is done by dividing the set of offspring in half (currently done by even replacement rank versus odd replacement rank), and using the rendezvous algorithm to get the even ranked offspring to point to the odd ranked offspring, and vice-versa. With this set of pointers, we can now complete the Crossover. Generate a random number between 0 and the length of the chromosome (i.e., three times the length of a message, since the chromosome contains both of the conditions and the action) for each even ranked offspring, to be used as the crossover point. Now the even ranked offspring **Get** the odd ranked offspring's chromosome (conditions and action). Now, using parallel load-byte and deposite-byte, swap the high order bits of the chromosomes up to the crossover point. Finally, store the crossed chromosomes back in the even ranked offspring and **Send** the crossed chromosomes to the odd ranked offspring.

10.5.3 Mutation

The Mutation algorithm uses a Poisson distribution to mutate zero, one, two, or three genes in the chromosome (bits in the conditions and action) of the new offspring. The Poisson distribution is implemented in parallel using a table (see Wagner 1969). A random number between 0 and 1000 is generated for each offspring. A match to the Poisson table indicates how many mutations to make (normally, very few mutations are done). For each mutation, a random bit position is picked, and a random new value (one, zero, or don't care) is set in that position.

10.5.4 Summary: Parallel Genetic Algorithms

To summarize, the parallel Genetic Algorithms used in *CFS are all independent of Classifier population size (up to 65,536). The total time for the Genetic Algorithms is around 400 milliseconds (or an average of 36 milliseconds per cycle in the experiment described above, since they were run once every 11 cycles). Very little optimization has been done on this part of the system, so the figure can probably be reduced by a factor of two or three. All these algorithms make heavy use of the communications mechanisms in the Connection Machine, both explicitly (with uses of **Send** and **Get** in parent copying and crossover) and implicitly (with uses of **Enumerate** and **Rank**). Two crucial algorithms that support the Genetic Algorithms are parallel random weighted selection (for picking parents and offspring) and parallel rendezvous (to establish pointers between parents and offspring and between pairs of mates in crossover).

10.6 Summary

The implementation of a parallel version of Genetic Algorithms in a Classifier System on a fine-grained massively parallel computer, the Connection Machine, provides verification that these inherently parallel algorithms can, in fact, be fully parallelized on a computer. This is best demonstrated by the speed of the system, which is independent of the number of Classifiers. The speed of those parts of *CFS other than its Genetic Algorithms depends only on the size of the message list and is linear.

This work also provides a demonstration of the power and importance of data parallelism as a programming style. It took less than one man-month to develop the first working version of the basic *CFS system, starting only with a description of CFS-C (the documentation in Riolo 1986a, Riolo 1986b). This also provides confirmation that the Connection Machine system is well-adapted to data parallelism and is easy to program.

The Learning Laboratory described here provides a framework for evaluating different approaches to machine learning on the same or similar tasks. This will aid future exploration and evaluation of new parallel Genetic Algorithms in Classifier Systems, and will also provide a way to compare and contrast Genetic Algorithms and Classifier Systems with other approaches to machine learning.

Finally, parallel implementations of Classifier Systems and Genetic Algorithms, like *CFS, provide a vehicle for exploring large task domains. Ultimately, a Classifier System with a large population should be able to evolve a set of rules for any repetitive structured task for which an evaluation function can be defined. An example of a difficult task that is unlikely to ever work with a small population of Classifiers, but might work with a large population, is the game of Go. The best current Go playing program, by Wilcox (Wilcox 1985), is only slightly better than a novice. Can we express the knowledge embedded in Wilcox's program in a set of rules and an evaluation function, and use that as the starting point for evolving rules for a really good Go player? Parallel implementations of Classifier Systems and Genetic Algorithms allow us to begin investigating such questions.

Bart Selman and Graeme Hirst

Chapter 11
Parsing as an Energy Minimization Problem

Abstract

We show how parsing can be formulated as an energy minimization problem. We describe our model in terms of a connectionist network. In this network we use a parallel, stochastic relaxation scheme similar to the one used in the Boltzmann machine (Fahlman, Hinton and Sejnowski 1983) and apply simulated annealing (Kirkpatrick, Gelatt and Vecchi 1983). We show that at low temperatures the time average of the visited states at thermal equilibrium represents the correct parse of the input sentence.

Our treatment of parsing as an energy minimization problem enables us to formulate general rules for the setting of weights and thresholds in the network. We also consider an alternative to the commonly used energy function. Using this new function, one can choose a provably correct set of weights and thresholds for the network and still have an acceptable rate of convergence in the annealing process.

The parsing scheme is built from a small set of *connectionist primitives* that represent the grammar rules. These primitives are linked together using pairs of computing units that behave like discrete switches. These units are used as binders between concepts represented in the network. They can be linked in such a way that individual rules can be selected from a collection of rules, and are very useful in the construction of connectionist schemes for any form of rule-based processing.

11.1 Introduction

Recently, we have seen a strong interest in the application of stochastic relaxation techniques using simulated annealing (Kirkpatrick *et al* 1983, Fahlman *et al* 1983, Geman and Geman 1984). These techniques are well suited for implementation on massively parallel hardware and their behavior can be characterized in terms of an energy minimization problem. These properties have led to an effort to search for reformulations of traditional serial algorithms into energy minimization problems. For example, Ballard (1986) shows how a restricted form of unit-clause resolution theorem proving can be reformulated into a relaxation problem. And Hopfield and Tank (1985) obtain promising results with the reformulation of the traveling salesman problem into an energy minimization problem[1].

[1] Note that Hopfield and Tank do not apply simulated annealing; instead they use an analog computation to find a minimum (not necessarily global) in the energy function.

Here we will show how parsing can be handled in such a framework.

We will formulate our work in terms of a connectionist network (Feldman and Ballard 1982). Such networks, consisting of large numbers of relatively simple processors, can be viewed as a natural implementation of a parallel relaxation algorithm. In particular, we will rely on a special class of these networks called Boltzmann machines (Fahlman *et al* 1983). The underlying computational mechanism of the Boltzmann machine is a stochastic relaxation process using simulated annealing.

This work is part of the much broader study of the application of connectionist models in natural language understanding (NLU). Recently, several of such models have been proposed; for example, Waltz and Pollack (1985) and Cottrell and Small (1983) give models for word-sense and syntactic disambiguation, and Reilly (1984) gives a scheme for anaphora resolution. The models are based on the deterministic connectionist scheme (McClelland and Rumelhart 1981; Feldman and Ballard 1982). A central aspect of these schemes is that they process the different sources of knowledge used in NLU, such as lexical and world knowledge, in a highly integrated way; for example, the syntactic and semantic processing are combined.

A major limitation of these schemes is their very limited capability to handle tasks such as parsing and case filling which seem to require processing to be based on a set of rules. For example, Waltz and Pollack use the output of a conventional chart parser to generate a network for the syntactic parse of the sentence. This network represents only the parse tree (or trees, in case of syntactic ambiguity) of the particular input sentence. Our approach will be more general, namely a network that directly represents the grammar rules and is to be used for parsing of a large number of sentences[2]. A similar approach could be used for other types of rule-based processing, like case filling.

11.2 The model

We will cast, as mentioned above, our approach in terms of the Boltzmann formalism[3]. In this formalism, the behavior of the system during relaxation can be described as a search for a global minimum in the energy of the network

$$E = \sum_k E_{loc,k} \tag{1}$$

in which

$$E_{loc,k} = \left(-1/2 \sum_j w_{kj} s_j + \theta_k \right) s_k \tag{2}$$

[2]Related to our approach is a model proposed by Fanty (1985) based on a deterministic connectionist model.

[3]Note that, although our formulation will use terms from the connectionist paradigm, like computing unit, it could just as well be described in terms of a minimization problem of a multi-variable energy function. Feldman (1985) discusses various connectionist schemes based on the notion of energy minimization.

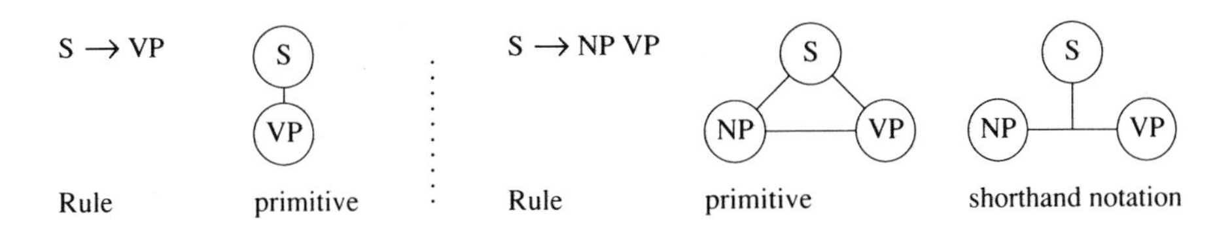

Figure 11-1: Two examples of connectionist primitives and their associated grammar rules (— excitatory link).

is the contribution of the kth unit with output value s_k and threshold θ_k to the energy, w_{kj} is the weight on the connection between the kth and the jth unit (we assume symmetrical connections), and the summation in equations (1) and (2) is over all units in the network. We will first discuss what we take our units to stand for and how these units are linked together (*i.e.* which of the links have non-zero weights).

11.2.1 Topology

We base our scheme on a context-free grammar, but this is not essential in our approach. The syntactic categories of the grammar are represented in a localist manner, that is, each syntactic category is represented by a unit in the network. As we will see, this localist approach allows us to represent the grammar rules in a very straightforward manner and consequently determines the set of non-zero weights (giving the topology of the network). The grammar rules will determine how these units are interconnected.

In the scheme we distinguish two layers. The *input layer* consists of a number of computing units representing the terminal symbols of the grammar. An input sentence will activate some subset of these units. Connected to this input layer is a network that represents the parse trees of all non-terminal strings in the language whose length is not greater than the number of units in the input layer. This network, called the *parsing layer*, is constructed from *connectionist primitives*, which represent the context-free grammar rules. Figure 11-1 gives two examples of such primitives. The activation of all units of a primitive corresponds to the use of the associated grammar rule in the parse. The number of units in the parsing layer depends on the particular context-free grammar rules and the number of input units. The different parse trees are not represented by completely disjoint sets of units, but share common substructures. This will keep the size of the network manageable.

We use intermediate computing units to link the primitives together. These units play the role of binders in the network and are therefore called *binder units*[4]. These units provide a lateral inhibition mechanism for selection of primitives. The computing units

[4]Units with a similar function are used by Cottrell (1985) and Fanty (1985) in their work on connectionist parsing using a deterministic model.

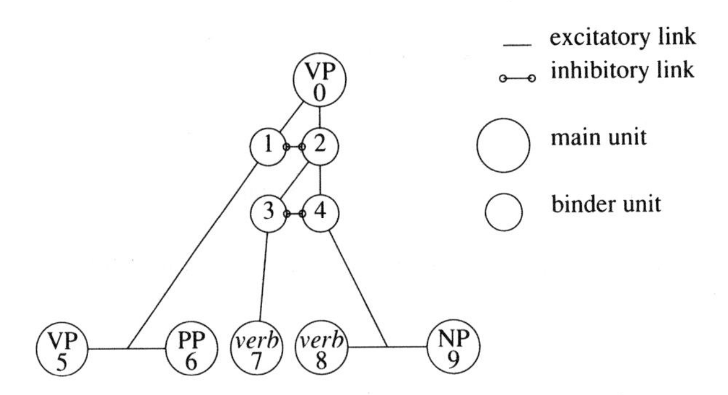

Figure 11-2: An example of the use of binder units (units 1, 2, 3 and 4).

representing the terminals and variables of the grammar are called *main units*.

Figure 11-2 gives an example of the use of binder units. The four binder units are used to represent the fact that the main unit #0 is part of three grammar rules:

$$
\begin{array}{lcll}
VP & \rightarrow & VP\ PP & (3a) \\
VP & \rightarrow & verb & (3b) \\
VP & \rightarrow & verb\ NP & (3c)
\end{array}
$$

The binders are linked in such a way, using inhibitory and excitatory connections, that when the network reaches a stable state the active binders (those whose output equals +1) tell us which one of the three possible grammar rules is used in the parse of the input sentence to decompose the verb phrase represented by unit #0. So, if binder #1 stays active, rule 3a is used in the parse; if binders #2 and #3 stay active, rule 3b is used, and if binders #2 and #4 stay active rule 3c is used.

11.2.2 Computational scheme

In this section we will consider the way in which the network finds the parse of a sentence. In the input layer of the network, the units are placed in *input groups*. Each group contains a unit for each terminal symbol of the grammar. The input groups are numbered; the nth group is associated with the nth word in the input sentence. Initially the computing units of both the input and the parsing layer of the network are inactive (their output is -1). As a sentence comes in, each word of the sentence activates the computing unit or units representing its associated syntactic category or categories. So the first word of the sentence activates one or more units (depending on the number of syntactic categories associated with the word) in input group #1, the second word one or more units in input group #2, and so on. After receiving input data, the network starts the relaxation process. During this process, the outputs of the activated computation units in the input groups are fixed at +1, while the outputs of other units in the input layer are fixed at -1, so that the network can find the optimal match between the input data and the internal constraints representing the grammar rules; this match will represent the correct parse of the input.

In our model we use a variation on the computational scheme of the Boltzmann machine (Fahlman, Hinton and Sejnowski 1983; Hinton and Sejnowski 1983, 1986), and apply simulated annealing (Kirkpatrick *et al* 1983) to find the optimal match between input data and internal constraints. Our scheme differs from the original in that we use -1 and $+1$ as output values of our computing units instead of 0 and $+1$. During the simulated annealing, the output of a unit k is set to $+1$ with a probability given by $(1 + e^{-\Delta E_k/T})^{-1}$, in which

$$\Delta E_k = 2(\sum_i w_{ki} s_i - \theta_k)$$

(*i.e.*, the difference in energy between a state with $s_k = -1$ and $s_k = +1$), and T is a formal parameter called the temperature. The use of $+1$ and -1 facilitates the representation of symmetrical interdependency relations between hypotheses in the scheme; there exists a one-to-one mapping between this scheme and the original (Selman 1985).

The fact that this scheme searches for a global energy minimum and that at equilibrium the relative probability of a particular state of the system is given by its energy enables us to formulate general rules for the setting of the weights on the connections and the thresholds of the computing units.

We compute the average value of the output of each unit at the different temperatures used in the annealing scheme. In the example given below, we will see how these average values will change during cooling of the system; finally, at a temperature just above the freezing point of the system, the units with outputs close to $+1$ will represent the parse of the sentence. To find the temperature just above the freezing point of the network, we consider statistical data on the behavior of the network during simulated annealing.

11.2.3 The setting of weights and thresholds

The setting of weights and thresholds is probably the most difficult problem in the design of a connectionist scheme. The set of weights and thresholds represents the internal constraints and therefore the knowledge in the system. So far we have described how units are interconnected in our parsing scheme; that is the set of links with non-zero weights. In this section we will discuss what values should be chosen for the weights on these links.

Given the fact that the network searches for a global energy minimum, we can, to a first approximation, analyze the behavior of the network by assuming that each unit and its direct neighbors will choose output values such that $E_{loc,k}$ in equation (1) becomes minimal. However this method gives only a rough approximation of the actual behavior, because minimizing E_{loc} for one particular unit often conflicts with minimizing

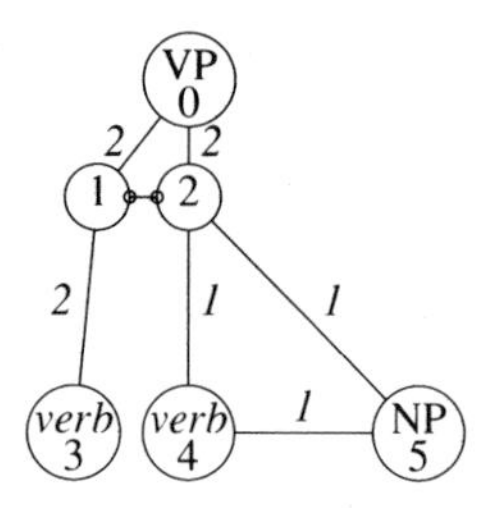

Figure 11-3: Some excitatory links in a typical configuration in the parsing network; the weights are given alongside the links.

E_{loc} of other units. To get a better insight in the behavior of the system we therefore consider the contribution to the global energy of small groups of units[5]. Because of the homogeneous structure of our network, we only have to consider a limited number of cases. As an example we will consider the setting of the weights on the excitatory links.

Figure 11-3 shows some excitatory links in a typical configuration. The network represents two grammar rules:

$$VP \rightarrow verb \qquad (4a)$$
$$VP \rightarrow verb\ NP \qquad (4b)$$

Rule 4a is represented by the units 0, 1, and 3; rule 4b by the units 0, 2, 4, and 5. During the relaxation process our network has to decide between rule 4a and rule 4b or neither of them. There is no a priori preference for one rule over the other. Therefore, because unit #2 is connected to two other units representing the left-hand side of grammar rule 4b and unit #1 is connected to only one unit representing the left-hand side of grammar rule 4a, we have to make the weight on the link between units #1 and #3 twice as strong as the links between units #2 and #4 and between units #2 and #5. (This can be easily generalized for grammar rules with more symbols; one chooses the weights such that the sum of the inputs at the binder units is equal for all grammar rules.) So we choose $w_{2,4}$ and $w_{2,5}$ equal to some positive constant and we set $w_{1,3}$ to twice this constant. We set this constant to 1.0. One should note that the absolute value of the constant is irrelevant. This value is only going to to determine at what temperature in our simulated annealing scheme the system is going to freeze, but the temperature is only a formal parameter introduced for simulated annealing and has no meaning in our final result.

For the use of a grammar rule in the parse, the presence of each symbol in the rule is equally important, and therefore we connect the units in a connectionist primitive representing a grammar rule with links of equal strength, so $w_{4,5} = 1.0$. And finally, because

[5]Of course, for an exact analysis one would have to consider all possible states of the (total) network; this becomes clearly infeasible for networks with more than about 25 nodes.

bottom-up and top-down parsing are completely integrated and of equal importance in our networks, we choose $w_{0,1} = w_{0,2} = 2.0$.

Selman (1985) gives similar analyses that lead to rules for the setting of weights on inhibitory links and of thresholds of the units. Here is a summary of these rules:[6]

$weight_{excitatory\ link}$	+1.0	*in primitive with three units*	(5a)
	+2.0	*in primitive with two units*	(5b)
$weight_{inhibitory\ link}$	−3.0		(5c)
threshold	0.0	*in main unit*	(5d)
	−2.0	*in main unit in symmetrical environment*	(5e)
	+2.0	*in binder unit*	(5f)

A *main unit in a symmetrical environment* is a main unit linked only to pairs of binder units (that is connected to both binders) and at most one other binder unit.

11.2.4 An alternative energy function

The rules for setting weights and thresholds given in the previous section are based on local properties of the network. Although good simulation results show that these rules have a high probability of leading to correct global behavior[7], they will not *guarantee* that the global minimum of the energy function will correspond to a correct parse. This problem is caused by the fact that admissible states[8] of a connectionist primitive can have different energies. This becomes clear, when we consider a state α of the network that corresponds to a correct parse (*i.e.* all primitives and binder pairs are in admissible states) and a state β with at least one primitive in a non-admissible state. Now, consider the energy change when going from state α to β; the increase in energy caused by the primitive(s) switching to the non-admissible state(s) can be offset by a decrease in energy of the other primitives changing to different admissible states. Therefore, state β can potentially have a lower energy than state α.

A possible solution to this problem is using sets of weights of different orders of magnitude (Ballard 1986, Derthick 1986). This corresponds to choosing the minimum energy gap between an admissible and a non-admissible state larger than the maximum possible change in energy caused by the total effect of primitives switching between admissible states. This approach has an important practical drawback; the number of different temperatures to be used in the simulated annealing process easily exceeds

[6] Often only the ratios between the different weights and thresholds are relevant; Selman (1985) gives these rules in a more general form.

[7] This is most presumably a consequence of the highly homogeneous structure of our parsing scheme (the networks are built from a small number of primitives).

[8] We refer to states of a connectionist primitive (or binder pair) that can be part of some parse tree as *admissible states*.

practical limits[9]. This follows from a theorem given by Geman and Geman (1984), which gives a sequence of temperatures that will guarantee the simulated annealing process to converge to the global energy minimum[10].

In this section we will consider an alternative approach to this problem, based on the choice of a different energy function. We define a new energy function E' as follows

$$E' = \sum_k E'_{loc,k} \tag{6}$$

in which

$$E'_{loc,k} = \begin{cases} +1 & \text{iff } E_{loc,k} \geq 0 \\ -1 & \text{otherwise.} \end{cases} \tag{7}$$

And we introduce the following terminology. The units in a connectionist primitive are *stable* if and only if all the primitive's units are on (output is $+1$), or all its units are off (output -1). The units in a pair of binder units are *stable* if and only if one of them is on when their parent[11] unit is on and both are off when the parent unit is off. Using this terminology, it follows that a correct parse corresponds to a state of the network in which all units are stable. Now using the alternative energy function, in which units contribute either $+1$ or -1 to the energy of the network, we can choose the setting of weights and thresholds in our network (Selman 1985) such that when a unit is stable it contributes -1 to the energy, and otherwise it contributes $+1$. Therefore, a correct parse will correspond to a state of the network with the minimal possible energy, $-N$. Any state with a higher energy will contain at least one unstable unit and therefore will not correspond to a correct parse[12]. The weights and thresholds used with the alternative energy function are of the same order of magnitude as those given in equation (5), and we obtain acceptable convergence properties in our simulated annealing scheme. This is confirmed by our simulation results. A drawback of our alternative energy function is that computing the change in the global energy when switching the state of a particular unit, as required during annealing, cannot be determined only by looking at the nearest neighbors of the unit; one also has to take into account the states of the next-nearest neighbors.

[9]Therefore, Derthick (1986) does not use the provably correct set of weights for his simulations. Ballard (1986) deals with correctness per se; he does not describe simulation experiments.

[10]Note that their result only gives an upper bound on the necessary length of the sequence of temperatures. The lower bound might be smaller; however Geman and Geman's bound is the smallest known.

[11]A pair of binder units always is linked to some 'parent'unit; see, for example, Figure 11-2.

[12]Basically, we have eliminated the problem that connectionist primitives in admissible states can contribute differently to the energy depending on their surrounding.

11.3 The design and testing of a network

To illustrate our model we will now consider an example. This network is based on the following context-free grammar rules (taken from an example in Winograd 1983):

$$
\begin{array}{rcl@{\qquad}rcl}
S & \rightarrow & NP\ VP & NP & \rightarrow & determiner\ NP2 \\
S & \rightarrow & VP & NP & \rightarrow & NP2 \\
VP & \rightarrow & verb & NP & \rightarrow & NP\ PP \\
VP & \rightarrow & verb\ NP & NP2 & \rightarrow & noun \\
VP & \rightarrow & VP\ PP & NP2 & \rightarrow & adjective\ NP2 \\
PP & \rightarrow & preposition\ NP & & &
\end{array} \tag{8}
$$

We will represent five input groups. In a complete network each input group has a unit for all terminals of the grammar; however to make our example network less complex, we will not represent each terminal in each input group.

For the grammar rules in (8), we can construct connectionist primitives similar to those given in Figure 11-1. To build the parsing layer upon the input layer, using these primitives, we consider the different possible ways in which the syntactic categories can be grouped according to the grammar rules, and design a network that represents those possibilities. One way we could have proceeded is by designing a set of networks, each representing the parse of one unique input sentence, linking all these networks to the input layer, and placing inhibitory connections between them. These inhibitory connections should guarantee that after the parsing network is given an input sentence, only the sub-network representing the parse of the input will remain active. However, many parse trees have common sub-structures. So one can save computing units by representing a common sub-structure by one set of units and linking that structure, using binder units, to the different parse trees represented in the network.

So, instead of constructing a network from a set of separate networks, each representing the parse of a sentence, we take an approach in which we try to share common syntactic structures between parse trees and minimize the number of main units. Following these guidelines we can construct from the input layer, using the connectionist primitives, a network like the one given in Figure 11-4[13]. The weights and thresholds in this scheme are set according to (5).

To demonstrate the parsing capability of our network we consider the input sentence[14]

$$
noun\ verb\ preposition\ determiner\ noun \tag{9}
$$

[13]Some simple input sentences show that the multiple units for *NP*s, *NP2*s, and *PP*s are necessary, but one could further minimize the number of *VP*s. However, this results in a network where the connectionist primitives are less visible, and which is therefore harder to understand.

[14]The network has been successfully tested for a number of input sentences, including some cases of syntactic ambiguity (in such cases more than one unit is activated within an input group), in which no semantic knowledge is necessary to resolve the ambiguity (Selman 1985).

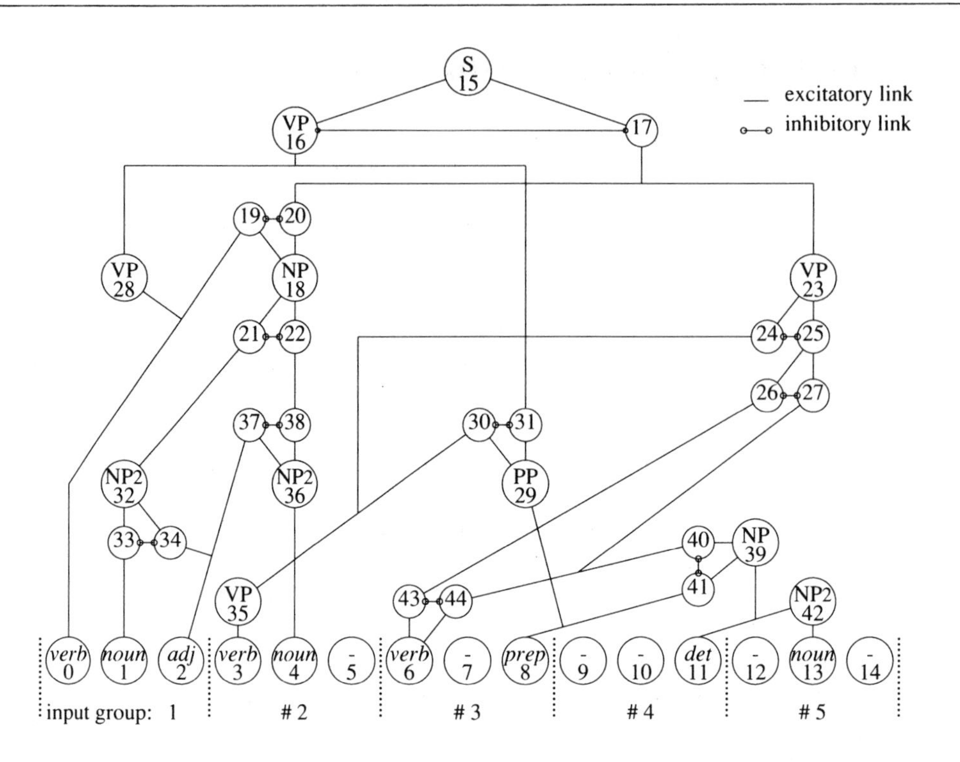

Figure 11-4: An example parsing network based on the grammar rules given in (8). Input group # 1 consists of units 0, 1 and 2; group # 2 consists of units 3, 4 and 5, and so forth.

exemplified by "*John ran down the hill*". For this sentence we ran a simulation of the parallel network on a serial machine using a simulated annealing scheme. To apply this scheme, one has to choose a descending sequence of temperatures such that the system has a reasonable chance of finding the state with the global energy minimum. Therefore one starts at a high temperature and first cools rapidly; once the system approaches the freezing point (the point at which it settles down in a state with a local or a global energy minimum; in this state the temperature is too low to escape from this minimum) one should cool very slowly. As we will see, we don't have to freeze the system completely; the right parse of the input sentence is found at a temperature just above freezing.

At each temperature above the freezing point one has to take sufficient computation steps to allow the system to reach equilibrium at that temperature.

To be able to choose the sequence of temperatures and the number of computation steps we did some test runs with the network. An appropriate sequence of temperatures was determined by considering the number of changes in the output value of each computing unit and the energy distribution of the system at each temperature. On the basis

+1.0 1 3 8 11 13

0.0 15 16 17 18 19 20 21 22 23 24 25 26 27 28 29 30 31 32 33 34 35 36 37 38 39 40 41 42 43 44

−1.0 0 2 4 5 6 7 9 10 12 14

T = 10000

+1.0 1 3 8 11 13

42

15 32 33 35

39 41

0.0 16 17 18 20 21 24 28 30

22 23 31 38

19 25 26 27 40 43 44

34 36 37

−1.0 0 2 4 5 6 7 9 10 12 14

T = 4.0

+1.0 1 3 8 11 13 15 17 18 20 21 23 24 29 30 32 33 35 39 41 42

0.0

−1.0 0 2 4 5 6 7 9 10 12 14 16 19 22 25 26 27 28 31 34 36 37 38 40 43 44

T = 1.0

Figure 11-5: Average output values of computing units during simulated annealing. During the simulations the outputs of the units 0 to 14 were fixed to represent the input sentence.

of these indicators we chose a sequence of temperatures starting at $T = 10000$ (to randomize the system), followed by $T = 4.0$, $T = 2.0$, and then in steps of 0.2 down to $T = 0.6$.

To estimate the required number of computation steps at each temperature, we considered the results of a sequence of simulations in which this number was slowly increased. When the average output values of the units become independent of the number of computation steps we assume that enough computation steps have been taken to scan the energy distribution of the system at equilibrium. Two thousand computation steps (*i.e.* 2000 updates of each unit) per temperature appeared to be sufficient. It should be noted that we did not try to minimize the number of temperatures or the number of steps per temperature.

Figure 11-5 shows the annealing process for sentence (9) at three different temperatures. Each panel shows the average output value of each computing unit at the temperature given below the panel. The numbers 0 to 44 are the numbers of the computing units. The vertical position indicates their average output value on the interval $[-1, +1]$. At $T \approx 1.0$ the system freezes; below this temperature the system stays in one state. Comparison of these results with the parse tree of this sentence given in Figure 11-6 shows that the time average of the outputs of the units, at a low non-zero temperature, corresponds to the correct parse of the input sentence if one chooses the units with outputs close to $+1$ as being part of the parse tree. At low temperatures there is a clear distinction between units with output close to $+1$ and the other units, as can be seen in

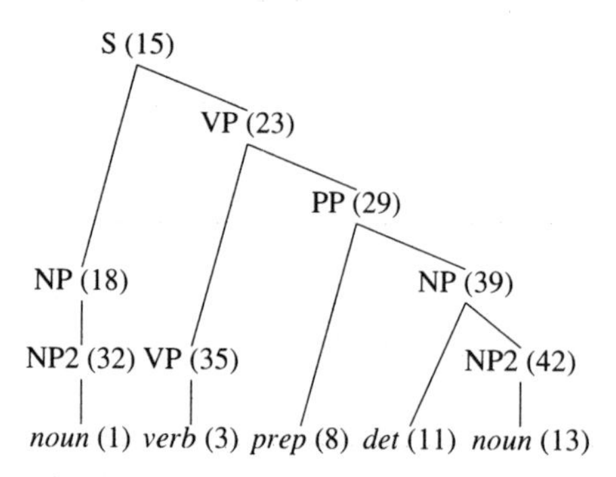

Figure 11-6: The parse tree of sentence (9). The numbers between parentheses are the numbers of the corresponding units in the parsing network.

Figure 11-5.

Although the set of average output values of the units in this section does not reveal any significant differences between the system in a frozen state or just above the freezing point, more information about the parse can be obtained at a temperature just above the freezing point, as we will see in the next section.

11.4 The robustness of our scheme

The weights in the example network given above were set in accordance with the rules given in (5). Because the setting of the weights and thresholds is a central issue in connectionist models, we will now consider the sensitivity of our scheme to changes in these values. We use again the example network given in Figure 11-4 and input sentence (9).

First we set the threshold of main unit #32 equal to zero; originally this threshold was set to -2.0, following rule (5e). The simulation results show that in the frozen state the system gives the correct parse, except for the main nodes #18 and #32, and the binder #21; that is the average output values of those units are -1.0. However the averages of the output values of these units at a temperature just above freezing are almost equal to zero. This means that at that temperature these units are part of the parse of the sentence for approximately 50% of the time (all other average output values are close to -1.0 and $+1.0$, consistent with the correct parse). This result can be explained as follows. Just above the freezing point the system jumps between two states, namely:

– state α, all units of the correct parse are active; and
– state β, same as state α, except for the units 18, 21, and 32.

States α and β have approximately the same low energy; however to jump between these states the system has to visit a state with a higher energy. At a temperature above the freezing temperature there is enough thermal energy to visit the intermediate state with a higher energy; in other words the system has a reasonable chance of visiting the intermediate state compared to the chance of being in the lower energy states α and β. Therefore, the system will jump between states α and β and the average output values of the units 18, 21, and 32 will be around zero. However, when the temperature is lowered the system freezes, that is it will settle in one of the states α or β, and there is not enough thermal energy to jump to the other low-energy state.

This example clearly demonstrates that the average output values of the units give more relevant information when determined just above the freezing point of the system than at or below that point. It also demonstrates how the Boltzmann mechanism not only finds a global minimum in the energy, but just above the freezing point the system jumps between a number of states with energies close or equal to the global energy minimum. This is an important advantage over a deterministic scheme. Even when such a scheme manages to find one of the states α or β, it is very unlikely that a deterministic scheme, just before finding α or β, would pass through the other state with minimal energy. This example also shows that if we don't follow all the rules given in (5), the system does not behave as well as when we do; however the model still comes up with a result close to the correct parse.

We will now consider what happens if we increase both the strength of the inhibitory links between binder units and the thresholds of these units. We choose a weight of -20.0 on the inhibitory links and thresholds of $+19.0$. These values are in accordance with the general rule for setting the thresholds on binder units and the weights on inhibitory links between them. In this case we don't find any significant differences between the simulation results with this new choice of weights and thresholds and those using the original values.

This is an interesting result, because with this choice of weights it becomes extremely unlikely, at low temperatures, to find a pair of binder units with both outputs equal to $+1.0$. Such a pair would give a large positive contribution to the energy; see equation (2). Therefore, the pairs are functioning as three-state switches with at most one unit with output equal to $+1.0$. This is useful during the search for a correct parse (or global energy minimum), because a pair with both outputs equal to $+1$ corresponds to the obviously incorrect situation in which two grammar rules are applied at the same time to decompose a syntactic category.

In Figure 11-2 we saw two pairs of binder units linked in such a way that they can choose the application of one specific grammar rule out of three. Using a similar approach one can design a network from pairs of binder units that can select one rule out of a large collection; such networks will be useful in general connectionist schemes for rule-based processing.

11.5 Conclusions

The task of computing the parse tree of a sentence can be viewed as a constraint-satisfaction problem in which the constraints are given by a set of grammar rules. We have shown how this task can be formulated in terms of an energy minimization problem and can be solved using a stochastic parallel relaxation algorithm using simulated annealing.

We have presented our results in terms of a connectionist network. In the design of such a connectionist scheme, the use of pairs of intermediate units that function as binders between units that represent the different concepts appears to be very useful. One can choose the thresholds of the binder units and the weights on the inhibitory links in between them such that they function as three-state switches (both units on is a forbidden state). These pairs linked together in a binary tree structure can be used to select one rule (for example, a grammar rule) out of a collection of alternatives. During the search for an optimal match between input data and the internal constraints in the network, the binder pairs select different rules to test whether they should be used. Interestingly, this bears a close resemblance to how a sequential processing scheme tries rule after rule; the advantage of the connectionist scheme is that many rules, each part of a different collection and represented in different parts of the network, can be applied in parallel, and also there is a complete integration of bottom-up and top-down processing.

We have discussed several approaches to the setting of weights and thresholds in the parsing network. One approach is based on analyzing the energy of small groups of units and some general symmetry considerations. In practice this leads to a correct global behavior of the network. A problem with this approach is that in order to *guarantee* correctness one would have to introduce sets of weights of different orders of magnitude. This would lead in general to unacceptably slow convergence to the minimum energy state in the simulated annealing scheme. We discussed an alternative energy function (equations 6 and 7) as a possible solution to this problem.

Finally, we have seen how in the annealing scheme, the network at temperatures just above the freezing point visits a number of states with energies equal or close to the global energy minimum of the network. Such states will, in general, show only minor differences from the state of the network that represents the correct parse; this makes the network less dependent on the particular choice of weights and thresholds.

The next logical step in this research is the addition of a semantic component in our scheme, to extend the disambiguation capability. Such a model would incorporate rules for case filling.

Acknowledgments

This work was supported by a Government of Canada Award to the first author, and grants from the University of Toronto and the Natural Sciences and Engineering Research Council of Canada to the second author. We would like to thank Dana Ballard, Garrison Cottrell, Jerry Feldman, Geoffrey Hinton, and Mike Luby for useful discussions and comments, and Sue Becker and Gail Godbeer for proofreading this paper. Thanks are also extended to the referee for helpful comments.

David S. Touretzky and Geoffrey E. Hinton

Chapter 12
Pattern Matching and Variable Binding in a Stochastic Neural Network

12.1 Introduction

DCPS, a Distributed Connectionist Production System, was developed as a demonstration that connectionist models could exhibit general symbol processing behavior (Touretzky and Hinton 1985, Touretzky and Hinton 1986). Connectionist models, also known as PDP (Parallel Distributed Processing) models, are networks composed of many simple homogeneous computing elements that are highly interconnected (Feldman 1982, Rumelhart and McClelland 1986). In designing these networks, connectionists take inspiration from what is known about the structure of the brain, but their models are not required to be anatomically or physiologically correct. The central idea is that from the collective activity of many individual units, each computing a simple function in parallel—as neurons do—intelligent behavior may emerge.

DCPS is a forward chaining production system architecture. It is similar in flavor to OPS5 (Brownston et al 1985), and not to systems like EMYCIN which employ goal-directed backward chaining (Davis and King 1977). DCPS uses highly restricted rule and working memory formats; it is not as powerful as OPS5 or EMYCIN, but in some respects it is quite powerful, particularly since it performs rule matching entirely in parallel.

The DCPS model is implemented as a Boltzmann machine: a stochastic Hopfield net (Hopfield 1982) that employs simulated annealing to search for global energy minima (Fahlman, Hinton, and Sejnowski 1983, Ackley, Hinton, and Sejnowski 1985). In this case, a global energy minimum corresponds to the selection of a production rule and a pair of working memory elements that match the rule's left hand side pattern. Rules contain variables on their left hand sides that constrain the match; variables may also appear on the right hand side, where their values are instantiated during rule firing. Pattern matching and variable binding—the heart of the rule match process—are both accomplished in DCPS by simulated annealing (Kirkpatrick, Gelatt, and Vecchi 1983).

In the following sections we give a brief introduction to Hopfield networks and Boltzmann machines and then describe the DCPS architecture in detail. The chapter concludes with some experimental observations about the model's behavior during annealing.

12.2 Hopfield Nets and Boltzmann Machines

A Hopfield net is a neural network model in which units have binary states and symmetric weighted connections, state changes occur asynchronously, and there are negligible transmission delays between units. Let α denote a state of the entire network, and let s_i^α denote the state of unit i when the network is in state α. Since units have binary states, s_i^α will be either 0 or 1. Let θ_i be the unit's threshold, and $w_{i,j}$ the real-valued weight between units i and j. Positive weights make excitatory connections, while negative weights make inhibitory ones. Note that $w_{i,j} = w_{j,i}$ because connections in a Hopfield net are symmetric. Self-recurring connections are not part of the model, so $w_{i,i} = 0$ for all i.

The input to a unit is the sum over all its connections of the weight on a connection multiplied by the state of the unit it connects to. We define the "energy gap" of a unit to be the amount by which its input exceeds its threshold. The reason for calling this quantity the energy gap will become apparent shortly. The energy gap of unit i when the network is in state α, written $\Delta E_i(\alpha)$, is given by

$$\Delta E_i(\alpha) = \sum_j s_j^\alpha \, w_{j,i} \; - \theta_i$$

Each unit continually checks its energy gap and sets its state to 1 if its input is above threshold, otherwise it sets its state to 0. A state change will of course affect the inputs to other units. Hopfield proved that such networks always settle into a stable state (which is often the stable state closest in Hamming distance to the starting state), because every time a unit changes state it reduces a certain global energy measure E. Stable states are therefore synonymous with minimum energy states. The energy of state α, written $E(\alpha)$, is the sum of the thresholds of the active units minus the sum of the weights between pairs of active units:

$$E(\alpha) = \sum_i s_i^\alpha \, \theta_i - \sum_{i<j} s_i^\alpha \, s_j^\alpha \, w_{i,j}$$

The energy gap of a unit is the decrease in the energy of the whole network caused by turning the unit on (given the current states of all other units).

Hopfield nets can easily get stuck in non-global minima. The Boltzmann machine, invented by Hinton and Sejnowski, uses stochastic binary state units which allow it to escape from such minima. In a Boltzmann machine a unit's energy gap determines the probability p_i that the unit will be in state 1 after an update. The probability is computed using the nonlinear function:

$$p_i(\alpha) = \frac{1}{1 + \exp(-\Delta E_i(\alpha)/T)}$$

The parameter T in the above equation is the "temperature." Figure 12-1 shows the probability p_i of a unit's being on as a function of its energy gap for three different

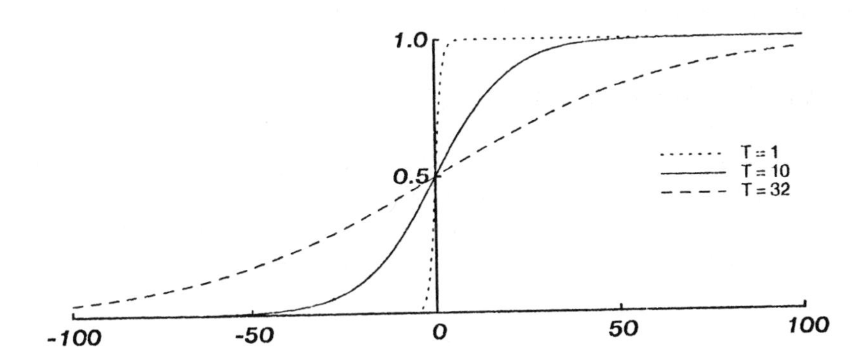

Figure 12-1: Probability p_i of a unit's being on as a function of its energy gap, graphed for three different temperature values. At low temperatures, the curve approximates a step function.

temperature values. At very high temperatures units behave randomly because p_i is roughly 0.5 no matter what the energy gap. At moderate temperatures the energy gap determines the *tendency* of a unit to come on: the higher ΔE_i, the more likely the unit is to be on. As the temperature drops, units behave more and more deterministically, until at absolute zero the units in a Boltzmann machine obey the same deterministic decision rule as in a Hopfield net.

If the network is run for a sufficiently long time at a non-zero temperature it approaches "thermal equilibrium" at this temperature. Even though the states of the units keep changing, the probabilities of global states obey the Boltzmann distribution in which the relative probability of any two global states depends only on their energy difference:

$$\frac{P(\alpha)}{P(\beta)} = \exp(-(E(\alpha) - E(\beta))/T)$$

Thermal equilibrium at a given temperature can generally be approached fastest by starting with a higher temperature and gradually reducing it—a process known as simulated annealing. An alternative to simulated annealing is to approximate the behaviour of the stochastic system by using real-valued units instead of binary ones. The value of a unit can be viewed as the probability of the unit being on, but the network explictly passes around these real-values (Hopfield and Tank 1985). Gradually increasing the gain of the deterministic, nonlinear I/O function of the units gives results similar to gradually decreasing the temperature during simulated annealing: The network tends to settle into the global energy minimum. This method generally runs faster on a conventional computer where there is little penalty for passing accurate real-values around.

12.3 An Overview of DCPS

The working memory of DCPS contains triples of symbols such as **(F A B)** or **(K V R)**, drawn from a twenty-five letter alphabet. The memory typically holds from two to a dozen triples at a time. A rule's left hand side consists of two clauses that specify triples, with each clause containing a variable in the first position; the variable must be bound to the same value in each clause in order for the rule to match. An example rule is:

```
(=x A B)  (=x C D)   -->  +(=x A D)  +(P =x Q)  -(R =x =x)
```

This rule can match the pair of triples **(F A B)** and **(F C D)**, or **(G A B)** and **(G C D)**, but it cannot match **(F A B)** and **(G C D)** because the variable binding constraint would be violated. If the rule fired with the variable **=x** bound to **F**, it would add the triples **(F A D)** and **(P F Q)** to working memory and delete the triple **(R F F)**.

A block diagram of DCPS is shown in Figure 12-2. It consists of five spaces labeled WM, C1, C2, Rule, and Bind. WM holds the contents of working memory: a set of triples represented as an activity pattern over a set of 2000 binary state units. The two *clause spaces*, labeled C1 and C2, select the triples from working memory that a rule will match; like WM space they contain 2000 units each. Rule space contains a clique of forty units for each production rule. Bind space's 333 units implement the variable binding constraint during rule matching.

After the network has completed a match, it fires the selected rule, meaning it performs whatever actions appear on the rule's right hand side. These actions update working memory by adding and deleting triples. Space does not permit a discussion of how updating is accomplished; the details are available elsewhere (Touretzky and Hinton 1985, Touretzky and Hinton in preparation).

An important simplifying assumption made early in the design of DCPS was that only one valid match would be possible at a time, *i.e.*, there would be a unique rule and variable binding that together constitute the only correct solution to the rule match problem. This assumption eliminated the need for a conflict resolution phase between rule matching and rule firing, but did not completely eliminate the potential for ambiguity in the match, as will be shown later.

12.4 A Distributed Working Memory

The contents of working memory in DCPS is represented as an activity pattern over a space of 2000 units. Each unit has a receptive field table such as the one in Figure 12-3.

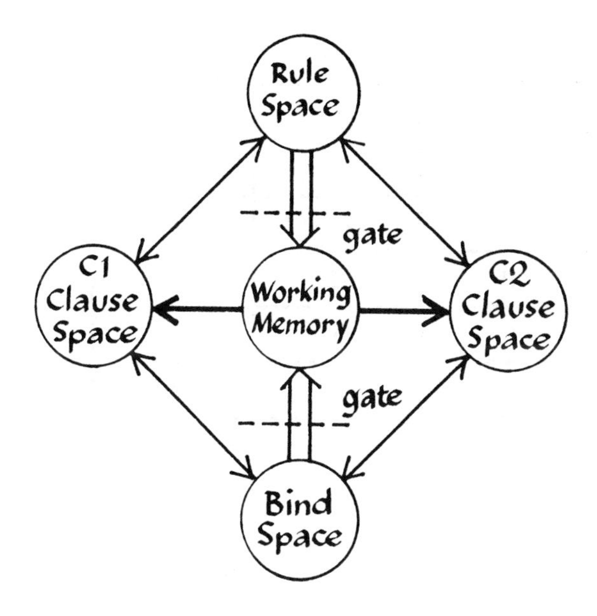

Figure 12-2: Block diagram of DCPS, a distributed connectionist production system.

A unit's receptive field is defined to be the cross product of the six symbols in each of the three columns, giving 6^3 or 216 triples. The unit described in Figure 12-3 has the triples **(C M R)** and **(F A B)** in its receptive field, along with 214 others. Receptive field tables are generated randomly prior to beginning the simulation; they determine the connection pattern between the units in C1 and C2 spaces and the units in Rule and Bind spaces. Once this wiring has been completed and the working memory units' states have been initialized, the tables are no longer needed; they are not consulted when running the model.

A triple may be stored in working memory by turning on all its receptors. With 2000 working memory units, triples will average $6^3/25^3 \times 2000$ or roughly 28 receptors; the number varies slightly from one triple to the next due to the random distribution of receptive fields. An external observer can test whether a particular triple is present in working memory by checking the percentage of active receptors for it. If this is close to 100%, the triple may be assumed to be present. For example, if the triple **(F A B)** were stored in working memory, the unit described in Figure 12-3 would be on, along with about 27 others. Although **(C M R)** also falls within the receptive field of this unit, the total number of receptors such unrelated triples have in common is small; on average, it is less than one. Thus, while 100% of the **(F A B)** units become active when **(F A B)** is stored, only 1 out of roughly 28 **(C M R)** units would become active. Clearly, **(F A**

C	A	B
F	D	G
H	G	J
P	M	L
S	Q	N
W	T	R

Figure 12-3: An example of a randomly-generated receptive field table for a working memory unit. The receptive field of the unit is defined as the cross product of the symbols in the three columns.

B) is present in working memory and **(C M R)** is not. The network itself never needs to compute these percentages, though. It relies on the fact that triples which are present have strong effects and triples which are absent do not.

Figure 12-4 shows the state of working memory when the two triples **(F A B)** and **(F C D)** have been stored. The 2000 working memory units are arranged in a 40×50 array, with the 55 that are active indicated by black squares. The positions of these 55 units in the array are not significant, since units' receptive fields are generated randomly. However, if we were to examine the receptive fields of each of the active units we would see that every one includes either **(F A B)** or **(F C D)**, or both.

Distributed representations such as this one, in which each item is represented by many units and each unit contributes to the representation of many potential items, have a number of interesting properties (Hinton, McClelland, and Rumelhart 1986). One of these is tolerance of noise. If after storing some triples in working memory a few units are flipped on or off at random, the perceived contents of working memory will not be affected at all.[1] Tolerance of noise is especially important when items will be deleted from the memory as well as added to it. A slight overlap in the receptor set of related triples causes deletion of a triple to affect any related ones previously stored. That is, if **(F A B)** and **(F C D)** were stored in memory and **(F A B)** were then deleted by turning off all its receptors, it is likely that only 27 of the 28 **(F C D)** receptors would

[1] Assuming, of course, that we do not require strictly 100% of a triple's receptors to be active for it to be considered present.

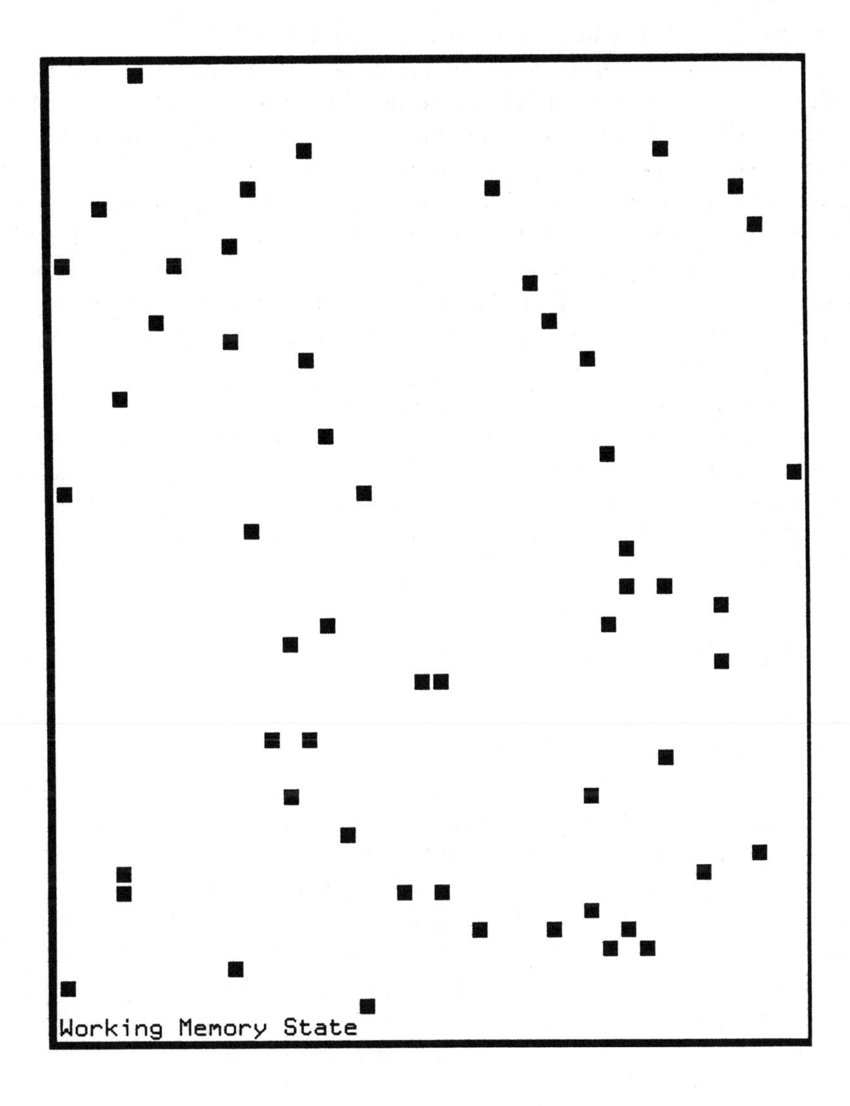

Figure 12-4: The state of working memory after the triples (**F A B**) and (**F C D**) have been stored. Active working memory units are indicated by black squares. 55 of the 2000 units are active.

remain active, the 28th having been shared between the two.

Another interesting property of the distributed representation is that the memory has no fixed capacity; instead its ability to distinguish stored items from other items decreases gradually as the number of stored items increases. Each triple added to working memory raises the number of active units, thereby increasing the support for other triples that have not been stored. As working memory fills up, the fraction of active receptors for certain triples that are "close" to those that have been stored approaches 100%, and the dividing line between present and absent triples blurs. If many closely related triples are stored, such as **(F A A)**, **(F A B)**, **(F A C)**, **(F A D)**, etc., then the system may exhibit local blurring, where it can't tell whether **(F A P)** is present or not but it is certain that **(G K Q)** is absent.

Finally, triples stored early on in a distributed memory eventually fade away if RHS actions delete a large number of other triples. This gradual decay phenomenon is again an effect of the overlap of receptive fields. One way to counteract the decay effect is to recall a triple before it has completely faded away, and then store it again. Whenever a triple is stored all its receptors become active, so its representation in working memory is refreshed.

12.5 The Structure of DCPS

Working memory units in DCPS act like latches during a rule match: they maintain their current state and supply the input that drives the rest of the system. At the center of the match process are the C1 and C2 clause spaces: their job is to each pull out one triple from working memory such that the pair of triples matches some rule's LHS while simultaneously satisfying the variable binding constraint that requires both triples to begin with the same letter.

A clause space works via a combination of inter-group excitation and intra-group lateral inhibition. There is a one-one mapping from WM space to C1 and C2 space; each WM unit tries to turn on its corresponding C1 and C2 units via an excitatory connection. However, mutually inhibitory connections among the units in each clause space allow only about 28 clause units per space to be active at a time—just enough to represent a single triple. Therefore, the patterns that appear in C1 and C2 space will be roughly 28-unit subsets of the current WM pattern. If these three spaces were run in isolation at any nonzero temperature, C1 and C2 would select continually-varying random subsets of the pattern in WM. However, due to the constraints imposed on C1 and C2 by connections from Rule and Bind spaces, the clause units that remain active at the end of an annealing will not be random: they will be the ones representing triples that best support a valid rule match.

Our clause spaces are instances of a generally useful device which Michael Mozer named the "pullout network" (Mozer 1984). A pullout network pulls out one pattern from a set of superimposed ones, subject to externally-imposed constraints. Mozer, who discovered the concept independently of its use in DCPS, proposed the pullout network as a way for a perceptual system to selectively attend to individual objects within its sensory field. Pullout networks have also been used in an architecture for manipulating symbol structures, called BoltzCONS, that is related to DCPS (Touretzky 1986a, Touretzky 1986b, Touretzky and Derthick 1987).

The Rule space of DCPS, which contains cliques of forty units per production rule, is organized as a winner-take-all network (Feldman and Ballard 1982). Units support other units in their clique via excitatory connections, and inhibit units in other cliques via inhibitory ones. The stable states of a winner-take-all network have all the units in one clique active and those in the remaining cliques inactive. By settling into one of its stable states, Rule space indicates which rule should fire. The choice of a particular stable state to settle into is influenced by activity in C1 and C2.

Each rule unit is wired up to C1 and C2 units according to the rule it represents. Bidirectional excitatory connections are built between the rule units and random subsets of the C1 units that code for triples matching the rule's first LHS clause; similarly, rule units are connected to C2 units that match the rule's second LHS clause. In the case of the example rule shown earlier, its rule units would connect to C1 units matching (**=x A B**) and C2 units matching (**=x C D**), for any value of **=x**. When triples such as (**F A B**) and (**F C D**) appear in C1 and C2 space and match a rule, they excite those rule units and cause Rule space to settle into the stable state representing that rule.

Bind space is organized as a *coarse-coded* winner-take-all network. There are twenty-five cliques, one per letter, and each clique has forty units, but each unit belongs to three cliques (votes for three letters) rather than just one. The space therefore contains a total of $(25 \times 40)/3 = 333$ units. Bind units connect to random subsets of the C1 and C2 units according to the letters they vote for. For example, a bind unit that votes for the letters F, J, and M would connect to C1 and C2 units where at least one of those letters appeared in the first column of the corresponding WM unit's receptive field table. Bind space has twenty-five stable states; in each of these, all the units that vote for one letter are active, and the remaining units are inactive.

Like Rule space, Bind space influences C1 and C2 via bidirectional excitatory connections; it constrains the two clause spaces so that they will pull out triples from WM that both begin with the same letter. Conversely, because the connections are bidirectional, whatever letter appears in the first position of the triples C1 and C2 spaces determines the stable state that Bind space settles into.

12.6 Matching by Simulated Annealing

WM, Rule, and Bind spaces all supply excitatory inputs to C1 and C2 to constrain the patterns that may appear there. The global energy minimum, which is reached when all these constraints are satisfied simultaneously, represents the selection of a rule and a pair of triples such that the two triples are present in working memory and match the rule's

1. **Initialize:** turn off all rule, bind, and clause units.
2. **Randomize:** run for 2 cycles at temperature 300. This temperature is high enough to ensure that all units which have any chance of being part of the solution have a reasonable chance of turning on, but it is low enough that completely irrelevant units are unlikely to be on.
3. **Match:** run for up to 10 cycles at temperature 32; stop if the energy is negative after any cycle.
4. **Cleanup:** run for 4 cycles at a temperature that is effectively zero. (We actually used 0.1 to avoid dividing by zero.)
5. **Rebias:** raise the threshold of all clause, rule, and bind units by 50.
6. **Verify:** run for 5 cycles at temperature of effectively zero.

Figure 12-5: The temperature schedule used in DCPS.

LHS. Unfortunately, there are many local energy minima generated by stable states of either the Rule or Bind spaces that do not constitute a valid match, because they leave at least one of the three major constraints unsatisfied. Therefore simulated annealing search would appear to be required to avoid local minima.

In practice, we have not had to use a genuine annealing search in order to get acceptable performance from the network. When we ran the network at zero temperature, it got trapped in poor local minima, but we discovered that this could be avoided by running at three distinct temperatures. Figure 12-5 shows the temperature schedule used in the current version of the model.

The network is initialized for matching by turning off all rule, bind, and clause units, leaving it in a zero energy state. Next its state is "randomized" by running it at a relatively high temperature of 300 for two cycles.[2] Units behave fairly randomly at this temperature, but they are still more likely to be active if their energy gap is positive than negative. At this temperature we have observed that the units that support the correct match and units that support partial matches are the ones that are on most often; units unrelated to a legal match become active less frequently. With so many units on, the energy of the network becomes quite high; with six rules (240 rule units) it varies between 8,000 and 12,000.

The real matching work is performed in the next step of the schedule, at a temperature of 32. The precipitous drop in temperature from 300 to 32 is more suggestive

[2] A cycle is N random updates, where N is the number of units in the network. Although the updating of units is done randomly, on average each unit will get one chance to update its state during each cycle.

of quenching than annealing but has no adverse effect on the match. The continued activity of rule, bind, and clause units now depends more strongly on support received from other units, but the network retains enough flexibility at this moderate temperature to explore various match possibilities rather than sink into the nearest local minimum. Cliques for a particular rule in Rule space or letter in Bind space may become very active, fade away, and become active again. Triples may materialize in the clause spaces, be partially replaced by other triples, and then perhaps return. At this temperature, thirty to sixty clause units at a time may be active in each clause space; the number is not limited to twenty-eight. The energy of the network rises and falls, but the general trend is decreasing. Once the energy falls below zero[3] the system is deep enough into a local minimum that it is unlikely to get out, so we move on to the cleanup step of the temperature schedule. In this step the network is run at a very low temperature, 0.1. Only units with positive energy gaps will remain active at this temperature. The result is that the clause spaces are left with roughly 28 units on, Rule and Bind spaces each have one clique active (40 units on), and the network is indicating as clearly as possible what it thinks the correct match should be.

12.7 Detecting Failed Matches

There are two ways in which the match can fail. The simplest is when the network fails to settle into a sufficiently low energy state. In this case very few of the units will have positive energy gaps, so when the temperature drops to 0.1 they will eventually all turn off. The more difficult case to detect is when the network has settled into a local energy minimum representing a partial match rather than the global minimum associated with the correct match. The energy of a partial match is moderately negative, typically around -2500. When the temperature drops to zero the network settles to the very bottom of the energy minimum and stays there.

All correct matches have energies below a certain value, which distinguishes them from partial matches. However, in connectionist models it is better if the behavior of inidividual units does not depend on measuring global properties of the network such as energy. To detect failed matches without measuring energy directly we use a technique called rebiasing. After the network has run for four cycles at a temperature of 0.1, the thresholds of all rule, bind, and clause units are raised by a value of 50, or equivalently, an inhibitory bias of -50 is applied to each unit. This has the effect of reshaping the energy landscape so that the correct match is still a deep energy minimum, but it is much narrower and its absolute energy is now considerably higher than zero. More importantly, a partial match that was a local minimum before is now located on a slope that leads down to the zero energy state with all units turned off. After rebiasing, the network is run for five more cycles at a temperature of 0.1. If units remain active at the

[3]This value is approximate and was determined empirically.

end of this step, the network is indicating a correct match. If a partial match was found, units will gradually turn off as a result of rebiasing, causing the energy to drop to zero.

One might wonder why the thresholds of the rule, bind, and clause units were not originally set at the higher level, eliminating the need for rebiasing. This would make the energy minima too narrow, making them more difficult for the search to find. Also, after rebiasing the energy of the correct match state becomes moderately positive. At high temperatures the network could find a better state simply by turning all its units off. When rebiasing is delayed until a low temperature has been reached, the network remains trapped in the state (now with positive energy, but still a local minimum) it was in if it managed to find the correct match.

12.8 Experimental Results

In an overnight simulation on a Symbolics 3600, DCPS ran a six-rule production system for more than a thousand successful rule firings. However, some match problems are more demanding than others, and situations can be contrived in which DCPS has difficulty finding the correct solution. Two such situations are discussed below.

In the simplest match cases there are no partial matches to worry about; the triples in working memory that do not match the winning rule do not match any of the other rules either. In more complex cases several feasible-looking matches exist with relatively low energy states; the system is forced to search among them to find the lowest one. This involves calling up different triples in the clause spaces for each possibility. As the number of partial matches increases DCPS becomes more likely to settle into a local minimum representing a partially successful match rather than finding the lowest energy state associated with the correct match. Figure 12-6 shows a set of rules and working memory elements that produce this behavior. In theory, annealing long enough and slowly enough would solve the problem, since the correct match is always a deeper energy minimum than any partial match.

In this match scenario there are six triples in working memory; the clause spaces must select from among the 36 possible ways to form a pair of triples the one combination the produces a correct match. What makes this problem difficult is the fact that four pairs of triples have fairly deep energy minima representing almost-sucessful partial matches. See Table 1. In these partial matches, either both clauses on the left hand side of rule Comb-1 or Comb-2 are satisfied but the variable binding constraint is not, or else only one of the left hand side clauses is satisfied but the variable binding constraint is met because both clause spaces support the same bind clique (J or K.) The source of the combinatorial confusion is the fact that all three rules and all three bind cliques are capable of getting full support from the clause spaces, so it's difficult to choose among them; what differentiates partial from complete matches is the fact that rule and bind space can't *both* get full support except when the rule is Comb-3 and the variable =**x** is

```
Rules:

Comb-1:    (=x A A) (=x B B)  -->  ...

Comb-2:    (=x C C) (=x D D)  -->  ...

Comb-3:    (=x E E) (=x F F)  -->  ...


Contents of working memory:
    (J A A)  (K B B)
    (K C C)  (J D D)
    (M E E)  (M F F)
```

Figure 12-6: A match situation in which combinatorial complexity hinders the search for a valid match.

bound to M.

DCPS does not search a combinatorial space by sequentially enumerating the possibilities. The partial representations of competing triples coexist simultaneously in the clause spaces, while rule and bind winner-take-all spaces host similar competitions. The stochastic nature of the Boltzmann machine causes some competitors in a space to fade out, and possibly fade back in again, until the network as a whole settles deeply enough into an energy minimum that a clear winner emerges in each space.

Figure 12-7 illustrates another contrived case where it is difficult for DCPS to conclude the match correctly. **(M J J)** is present in working memory but none of the rules Syn-1 through Syn-4 can match, due to their second clause. While all rules compete

Degree of Match	Triple in Clause 1	Triple in Clause 2	Rule Supported	Binding Supported
Partial	(J A A)	(K B B)	Comb-1	0.5 J, 0.5 K
Partial	(J A A)	(J D D)	0.5 Comb-1, 0.5 Comb-2	J
Partial	(K C C)	(J D D)	Comb-2	0.5 J, 0.5 K
Partial	(K C C)	(K B B)	0.5 Comb-1, 0.5 Comb-2	K
Complete	(M E E)	(M F F)	Comb-3	M

Table 12-1: The four partial matches generated by rules Comb-1 through Comb-3 have fairly deep energy minima, but there is a global minimum, representing the one complete match, in which all constraints are satisfied.

```
Rules:

Syn-1:    (=x J J) (=x A A)   -->   ...
Syn-2:    (=x J J) (=x B B)   -->   ...
Syn-3:    (=x J J) (=x C C)   -->   ...
Syn-4:    (=x J J) (=x D D)   -->   ...

Anti:     (=x R R) (=x S S)   -->   ...

Contents of working memory:
    (M J J)
    (M R R)
    (M S S)
```

Figure 12-7: A match situation in which synergistic action between four rules that generate partial matches prevents the system from finding the correct match.

with each other as a result of being in a winner-take-all network, the Syn rules also help each other by supporting **(M J J)** as the first clause. This synergy effect, which occurs whenever failing rules have related left hand sides, interferes with the search for the correct match. In order to find this match, the Anti rule must override the four Syn rules and get the pattern for **(M R R)** into C1 space. The more Syn rules there are to support **(M J J)**, the harder this will be.

12.9 Summary

DCPS treats pattern matching and variable binding as a parallel constraint satisfaction problem, and uses a degenerate form of simulated annealing to find a solution. The bulk of the work is performed at a single temperature. The model does not require gradual cooling to give acceptable performance, but it does rely on the stochastic behavior of units at the operating temperature of 32 in order to avoid getting trapped in local minima.

Decisions about the match result (*i.e.*, which rule matches and what the variable value should be) are made by two winner-take-all networks: the Rule and Bind spaces. The tendency of these networks to quickly settle into one of their many stable states must be counterbalanced by running at a sufficiently high temperature to ensure that DCPS tries different partial matches in search of the one complete match. For difficult match problems, the system must perform a combinatorial search through the space of possible rule choices and variable bindings. However, it does not consider the possibilities one at a time, as in some other annealing search applications. Instead, due to the use of distributed representations for triples, the activity patterns for competing working memory items are superimposed; triples may partially materialize in C1 and C2 spaces, influence the Rule and Bind spaces, fade out again, and possibly reappear. When a valid match is found, the energy of the network drops sharply and the system enters an energy well from which it cannot escape. The energy landscape during the match is such that there are many local minima associated with partial matches, but both they and the global minimum have broad, gently sloping sides. Rebiasing changes the energy landscape so that partial matches are no longer minima at all, but this also gives the global minimum steep sides, which would make it more difficult for the annealing search to find.

The use of Hopfield's energy measure to characterize incorrect matches is important because it allows the model to recover from settling into a local energy minimum when it fails to find the global one. The re-biasing step followed by continued annealing causes all the units to turn off if the current state is not the global minimum. Afterwards the match can be retried by running another annealing.

Acknowledgements

This work was supported by a grant from the System Development Foundation, and by National Science Foundation grants IST-8516330 and IST-8520359. We thank Scott Fahlman, Jay McClelland, David Rumelhart, and Terry Sejnowski for useful discussions, and Bruce Krulwich for contributing to the continued development of the simulation.

David H. Ackley

Chapter 13
An Empirical Study of Bit Vector Function Optimization

Abstract

Seven algorithms—including hillclimbers, variants of genetic search and simulated annealing, and a technique called *stochastic iterated genetic hillclimbing* (SIGH)—are tested on a set of optimization problems employing functions defined from a bit vector to the real numbers. The performances of the algorithms are measured by the average number of function evaluations required to find the optimal function value. No algorithm emerges as the unqualified winner, but as the functions become more complex, the annealing algorithm, the genetic algorithms, and SIGH perform better than simpler hillclimbing techniques.

The research reported here was performed as a part of dissertation research on a connectionist learning machine and search technique called SIGH, standing for *stochastic iterated genetic hillclimbing.* This paper is essentially Chapter 3 of my dissertation (Ackley, 1987), with occasional inclusions from other chapters. As such, many aspects of the problem formulation, the search techniques, and especially the motivation for SIGH receive scant coverage here. The results of a preliminary study, using fewer algorithms and a smaller set of test functions, have been reported in (Ackley, 1985). The reader is referred to the full dissertation for more information.

One way to measure the performance of a function optimization technique is to observe how good a function value the technique can find in a given amount of time. In this research, the converse approach is taken. Performance is measured by how many function evaluations are required before the technique finds a function value at least as good as a given value. The function-to-be-optimized is assumed to be in a "black box"—at the outset, a search technique is told only that the function takes a bit vector of a given dimensionality as input, and that it produces a scalar real value as output. The search strategy is not told the structure of the function or the range of possible function values. This problem poses two conflicting goals. On the one hand, a strategy must be capable of *learning while searching*: It must be able to gather global information about the space and use it to concentrate the search effort in the most promising regions. On the other hand, a strategy must be capable of *sustained exploration*: If focusing on the apparently most promising region does not lead to the discovery of an optimal point, the search strategy must be able to move on to other regions of the space and continue searching.

During a search, a hypothetical perfect strategy would quickly deduce the structure of the landscape, and adapt its behavior to find a solution state rapidly. To estimate how close any real search strategy comes to this sort of perfection, it is not enough simply to demonstrate that it can search an apparently difficult landscape quickly. After all, it might be that the assumptions made by the search strategy just happen to match the characteristics of the landscape. Such a demonstration does not allow one to infer anything general about the strategy's ability to learn while searching.

This paper presents results of simulations of a set of search strategies on a "suite" of test functions with qualitatively different characteristics. There are two primary goals. The first is to assess the generality of each search strategy with respect to the test suite. At the commencement of any given search, a strategy cannot know *which* test function it is facing. If a strategy turns in comparatively fast performances across the test suite, one may infer that, with respect to the test suite and the other algorithms, the search strategy is generally effective.

The second goal is to observe the behavior of the search strategies on the test functions, looking for insights into their strengths and weaknesses that may have applicability beyond just the functions in the test suite. There are a number of "rules of thumb" that are commonly believed about the abilities and drawbacks of various search strategies, and the second goal is to gain evidence in support of or against such conclusions. For example, it is sometimes held that the existence of local maxima in a space rules out hillclimbing approaches, but the results reported here suggest that conclusion is too strong.

The test suite used in this paper has six functions. It includes a linear function, two functions that each have one local maximum and one global maximum, a function with an exponential number of local maxima, a function that has large flat areas, and a function that combines the properties of the previous five. There are seven strategies tested: two uphill-only hillclimbers, a stochastic hillclimber, simulated annealing, two variations of genetic search, and stochastic iterated genetic hillclimbing. In all, seven algorithms are applied to six functions, each at four dimensionalities, with fifty simulation runs per strategy-problem-size combination, requiring a total of eight thousand four hundred individual searches.

This paper discusses the methodology behind the experiments, then presents the algorithms in enough detail that the interested reader should have little difficulty producing an implementation. The test functions are then considered one at a time. After motivating each function, simulation results are presented and discussed. Finally, the data from all the experiments are summarized.

13.1 Methodology

There are a number of difficulties inherent in this sort of experiment. At every turn, the cost of performing experiments limits the breadth and depth of the study. In many cases, the simulation results suggest further interesting experiments, but most such follow-up experiments will have to await future work. Furthermore, a major theme of this research is searching high-dimensional spaces. It would be fascinating to consider functions on hundreds or thousands of bits, but resource limitations, at present, rule this out. In general, even for "good" strategies, increasing the size of the problem increases the number of function evaluations required to find a solution. With over eight thousand searches required to simulate the cross-product of strategies by problems by sizes by runs, even modest-sized functions engender a significant computational burden. The first five functions are tested at 8, 12, 16, and 20 bits, corresponding to problem spaces with from 256 to 1,048,576 possible states. The final function is tested at 15, 20, 25, and 30 bits, corresponding to problem spaces with from 32,768 to 1,073,741,824 possible states.[1]

13.1.1 Notation

To present the algorithms uniformly, it is worthwhile to define some notation. In general, the scheme is to use a letter to indicate a type of object, with subscripts where necessary to suggest interpretations or to distinguish between different instances of a given type. The following conventions are used for all algorithm definitions

n = An integer representing the number of bits in the function space.

x = A bit vector of length n representing the location of a point in the function space.

f = A function from an n-bit vector into the finite reals representing the function-to-be-optimized.

v = A real-valued scalar representing the value of a point in the function space.

b = A single bit representing one dimension of a point in the function space. A superscript on a b indicates which point in the function space the bit is part of; a subscript indicates which bit within the point is denoted. For example, $(b_1^c, \ldots, b_n^c) = x_c$. Bit values are denoted by *0* and *1* to distinguish them from unit states.

p = A real-valued scalar in the range $(0, 1)$ representing a probability, e.g., the probability of a mutation in a genetic algorithm.

r = A real-valued scalar in the range $(0, 1)$ representing an exponential decay rate, e.g., the rate at which the temperature approaches a final value in a simulated annealing algorithm. A value of 1 indicates no decay; a value of zero indicates complete immediate decay.

t = An integer representing a moment in time, in terms of function evaluations.

[1] In (Ackley, 1987, Chapter 5), simulations are presented for test functions of 32 and 64 bits.

Some of the algorithms refer to the points that are "adjacent" to a given x; this denotes the set of points that are at a hamming distance of one from the given point, so that they differ from the given point on any one dimension.

Variable assignments are represented by "$\leftarrow$", e.g., $x_c \leftarrow x_a$ means that the current value of x_c is replaced by the current value of x_a. With two exceptions, everything represented by a symbol in this notation is assumed to be a variable. The first exception is parameters, which can be thought of as variables that are assigned to only once, before the algorithm begins, and are then treated as constants thereafter. The second exception is the function values, denoted by v's. Function values are viewed as dependent properties of x's, rather than as full-fledged independent variables. They remain attached to the point in the function space that generated them. When an x variable is assigned a new value, the corresponding v changes as well. For example, if $v_c = 10$ and $v_a = 20$, then after the assignment $x_c \leftarrow x_a$, $v_c = v_a = 20$.

13.1.2 Parameter tuning

Several of the search strategies include one or more adjustable parameters in their definitions. This presents another obstacle to performing a comparative study, since the efficiency of any given search strategy can usually be dramatically affected by the choice of parameter values, and the best parameter values for searching one function may be quite different from the best values for another function. For example, many algorithms have parameters that, in one way or another, affect the "learning rate" of the strategy. When such a strategy is tested on a linear function, the fastest performance is produced when large learning rates are chosen. However, on a more complex function, such "aggressive" learning rates often become a drawback, increasing the chance of wasting a lot of time on a local maximum, and producing a longer average search time compared to the speed possible with more moderate learning rate parameters.

To respect the black box problem formulation, I have assumed that each parameter must be given either (1) a constant value, (2) a value that depends on n, the dimensionality of the function space, or (3) a value that depends on other parameters. During exploratory simulations, I hand-tuned the parameters of each parameterized algorithm using the functions in the test suite. Since I was trying to assess the generality of the various algorithms, I attempted to find values for the parameters that yielded good performances on all the test functions. In most cases, I found that a change in a parameter value that cuts search time on one function will increase it on another function. I tried to balance out such effects to avoid terrible performances on any of the test functions, but that was not always possible. Some functions in the suite are essentially intractable for some of the strategies. Although I was sometimes able to find very different parameter settings that allowed strategies to succeed on those functions, those parameters usually produced a tremendous increase in search time on the rest of the functions.

It is perfectly possible that there are parameter values that would produce better average performance than the values I arrived at. Parameter tuning is more of an art than a craft, and it is important to keep this in mind when considering the simulation results. With different parameter settings, the ranking of the algorithms on any given test function might be changed. Nonetheless, in many cases there are good reasons—in terms of the assumptions a search strategy makes about function spaces—for expecting a given algorithm to do poorly (or well) on a given function, and in general it seems unlikely that parameter tuning can overcome such qualitative properties.

13.1.3 Non-termination

Although all of the algorithms tested are, in principle, capable of optimizing all possible functions, there is no strong bound on how long they may take to do it. Practically speaking, there is a risk that a given strategy on a given problem may be unable to find a solution state in any reasonable amount of time. It proved necessary to include a maximum time cut-off in the simulator to avoid the problem of a single search consuming excessive resources.

A limit of one million function evaluations was imposed for any single search. The limit worked as follows: If *any* of the fifty runs of a given strategy on a given problem at a given size exceeded one million evaluations, that search was aborted and the remaining runs of that strategy-problem-size combination are skipped. Such circumstances are designated in the tables by "$> 1M$," indicating that the time limit was exceeded for at least one of the fifty runs.

13.2 Seven algorithms

Here, then, is the list of algorithms tested in this paper. Each algorithm is given an acronym, some discussion of its operation, and a figure showing the details of the procedure and any adjustable parameters it may have.

13.2.1 Iterated hillclimbing–steepest ascent (IHC-SA)

Pick a random point, hillclimb along a path of steepest ascent until there are no uphill moves possible, and then start over. This very simple algorithm can be considered the iterative version of an even simpler one-shot hillclimber (HC). HC is not tested since it is only guaranteed to find the maximum in convex spaces. The IHC-SA algorithm appears in Figure 13.1.

IHC-SA is a "conventional" hillclimbing strategy. For a space of n dimensions, it requires n function evaluations to determine each move it makes. Even on a linear function, it will be expected to take $n^2/2$ evaluations to find the maximum. For large n, this may be unacceptably slow, and steepest ascent hillclimbing is therefore often discounted as unsuited to high-dimensional search problems. However, on the functions and sizes

Figure 13-1: IHC-SA: Iterated steepest ascent hillclimbing

1. *(Restart)* Select a point x_c at random and evaluate it.
2. *(Hillclimb)* Evaluate all points $x_1, \ldots, x_n$ that are adjacent to x_c, producing $v_1, \ldots, v_n$. Let v_u be the largest such v, and x_u be the corresponding point. If $v_u \leq v_c$, go to step 1. Otherwise let $x_c \leftarrow x_u$, and repeat this step.

Parameters: – none –

Figure 13-2: IHC-NA: Iterated next ascent hillclimbing

1. *(Restart)* Select a point x_c at random and evaluate it. Set $i \leftarrow 0$.
2. *(New current point)* Set $j \leftarrow i$.
3. *(Hillclimb)* Produce x_a from x_c by flipping the ith bit in x_c. Evaluate x_a. Set $i \leftarrow (i+1) \bmod n$. If $v_a > v_c$, set $x_c \leftarrow x_a$ and go to step 2. If $i = j$, go to step 1, otherwise repeat this step.

Parameters: – none –

considered in this paper, it is sometimes a strong competitor. In an empirical study, the power of hillclimbing should not be underestimated without actual comparisons.

13.2.2 Iterated hillclimbing–next ascent (IHC-NA)

The quadratic time complexity of steepest ascent hillclimbing does not carry over to all possible hillclimbing approaches. One way to cut down on the expense of evaluating all adjacent points before making a move is to try alternatives only until an uphill move found, and then to make that move without bothering to check whether there are other, better, moves possible. The IHC-NA algorithm considers adjacent points according to an arbitrary enumeration of the dimensions, and takes the first uphill move. The important point about this algorithm is that, *after* accepting a move, it doesn't return to the beginning of the enumeration and start over; instead, it tries the *next* dimension in the enumeration, and continues round-robin, looping back to the beginning of the enumeration when the last dimension has been tried. So, for example, if an uphill move is found on the fifth dimension in the enumeration, the first alternative checked after moving to

Figure 13-3: SHC: Stochastic hillclimbing

1. *(Initialize)* Select a point x_c at random and evaluate it.

2. *(Hillclimb)* Pick an adjacent point x_a at random and evaluate it. Select the new point (i.e., let $x_c \leftarrow x_a$) with probability $\frac{1}{1+e^{(v_a - v_c)/T}}$. Repeat this step.

Parameters

Symbol	*Range*	*Value*	*Description*
T	$T > 0$	10	The temperature

the new point is the sixth dimension in the enumeration. If the algorithm proceeds all the way around the list of dimensions and arrives back where it started without finding an uphill move, the current point must be a local maximum, and the algorithm restarts.

The IHC-NA algorithm has the nice feature that it is guaranteed to be optimal on linear spaces, requiring at most $n + 1$ function evaluations. In four of the six test functions, the next ascent approach performs faster than the steepest ascent approach. The algorithm appears in Figure 13.2.

13.2.3 Stochastic hillclimbing (SHC)

Stochastic hillclimbing is an alternative approach to avoiding the expense of enumerating all the adjacent points. It starts by picking a current point at random. Then it considers a random one bit change to the current point, and accepts the change (i.e., it selects the changed point as the new current point) with a probability determined by a fixed function of the difference in function value. This step is then iterated until the problem is solved.

Unlike the first two algorithms, SHC has no explicit provision for detecting when it is at a local maximum. Instead of detecting local maxima and restarting, SHC relies on its ability to accept downhill moves to escape local maxima. Moves that provide an improvement have a greater than 50% chance of being accepted, and moves to poorer positions have less. The greater the difference between the value of the current point and the value of the adjacent point being considered, the closer the probability comes to zero (for a very big downhill move) or to one (for a very big uphill move). The determination of "bigness" is performed by dividing the difference in function values by an adjustable parameter $T > 0$. Larger values of T cause the probabilities to be closer to 50-50, smaller values cause the probabilities to be closer to 0/1. The algorithm is given in Figure 13.3.

Figure 13-4: ISA: Iterated simulated annealing

1. *(Restart)* Set $T \leftarrow T_{max}$. Select a point x_c at random and evaluate it.
2. *(Stochastic hillclimb)* Pick an adjacent point x_a at random and evaluate it. Select the new point (i.e., set $x_c \leftarrow x_a$) with probability $\frac{1}{1+e^{(v_a - v_c)/T}}$. Repeat this step k times.
3. *(Anneal/Convergence test)* Set $T \leftarrow rT$. If $T \geq T_{min}$, go to step 2, otherwise go to step 1.

Parameters

Symbol	*Range*	*Value*	*Description*
T_{max}	$T_{max} > 0$	100	Starting temperature
T_{min}	$T_{max} > T_{min} > 0$	0.1	Minimum temperature
r	$0 \leq r < 1$	0.9	Temperature decay rate
k	$k > 0$	n	Time per temperature

13.2.4 Iterated simulated annealing (ISA)

In SHC, the "temperature" T of the system is a parameter, and is therefore held constant for the duration of a search. In the simulated annealing technique, the temperature is a variable that is started at a high value and is gradually reduced during the search. At high temperatures, the system accepts moves almost randomly, regardless of whether they are uphill or down. As the temperature is lowered, the probability of accepting downhill moves drops and the probability of accepting uphill moves rises. Eventually the system "freezes" in a locally or globally maximal state, and no further moves are accepted.

Since there is always the risk that the system may freeze without having found the global maximum, simple simulated annealing may fail to terminate in the black box computational model. Like iterated hillclimbing versus simple hillclimbing, however, it is easy to define an iterated version of simulated annealing that avoids this problem. Iterated simulated annealing performs a fixed number of function evaluations at each of a sequence of decreasing temperatures, and when the temperature drops below a minimum threshold, the algorithm restarts. The algorithm is given in Figure 13.4.

13.2.5 Iterated genetic search–Uniform combination (IGS-U)

To my knowledge, the specific forms of the genetic algorithms tested in this paper have not previously been studied, but their general form derives directly from the R1 genetic algorithm presented by Holland (1975). As discussed in (Ackley, 1987, Chapter 2), three basic components are required for a genetic algorithm: (1) a population of isomorphic structures, (2) a combination rule to produce new structures from old structures, and (3) some method for ensuring an exponential spread of high-valued schemata through the population.

In a "traditional" genetic algorithm, the third step is accomplished by the "reproduction with emphasis" operation, wherein extra copies of high-valued points are introduced into the population and low-valued points are deleted. Such an operation ensures that the change in the number of instances of a schema is proportional to the difference between the observed value of the schema and the average value of the population. Iterating the reproduction step, therefore, produces an exponential increase for the high-valued schemata (and an exponential decrease for the low-valued schemata.)

In the R1 algorithm, this reproduction with emphasis is accomplished incrementally, by a weighted probabilistic selection of "parents" from the population. If the value of a point is equal to the average value of the population, for example, that point has a $\frac{1}{m}$ chance of being selected as the first parent, where m is the size of the population. (It also has a $\frac{1}{m}$ chance of being selected as the second parent—there is generally no provision for ensuring that a point does not combine with itself.) If the value of a point is ten times greater than the average, that point has a $\frac{10}{m}$ chance, and so forth. After selecting two parents in this way, R1 applies various "genetic operators" such as crossover and mutation to generate a new point. This point is then added to the population. Since the population is kept at a fixed size, some other point must be displaced to make room for it. In the R1 algorithm, this is accomplished by a uniform random selection among all the members of the population. The new point is evaluated, the average value of the population is updated, and one iteration of the R1 algorithm is complete.

In the R1 algorithm, although reproduction opportunities occur in proportional to fitness, opportunities for "survival" (i.e., remaining in the population) do not. All members of the population are equally likely to be "killed" to make room for the new point. Consider this algorithm: Pick two parents from the population with a uniform probability distribution, combine them to produce a new point, and then make room for it in the population by deleting a member with probability in proportion to its "unfitness." In other words, pick a point to displace based on how far *below* the average value of the population it is. For example, a point with a value equal to the average value has a $\frac{1}{m}$ chance of dying, a point with a value half the average has a $\frac{2}{m}$ chance of dying, and a point with a value ten times the average has a $\frac{1}{10m}$ chance of dying.

Like the R1 algorithm, this "survival of the fittest" algorithm produces an exponential spread of high valued points. Each member gets a fair chance to reproduce on each iteration, and the number of iterations that a member is likely to survive is proportional to its fitness. Instead of employing "reproduction with emphasis," this algorithm does the job via "termination with prejudice." A small simplification of this algorithm yields the genetic algorithms that I employed in this study.

Figure 13-5: IGS-U: Iterated genetic search–Uniform combination

1. *(Restart)* Select a population of m points $x_1, \ldots, x_m$ at random and evaluate them. Compute the average value of a point: $\theta \leftarrow \frac{1}{m}\sum_{1 \le i \le m} v_i$. Set $t \leftarrow 0$. Set $p \leftarrow p_{max}$.

2. *(Crossover/Mutate)* Pick two points x_i, x_j from the population at random. Create a third point x_c as follows. For each bit $b_1^c, \ldots, b_n^c$ in x_c, with probability p, set its value to *0* or *1* at random, otherwise choose x_i or x_j at random and copy the corresponding bit value. Evaluate x_c.

3. *(Reproduce)* Pick a point x_d from the population at random subject to the constraint that $v_d \le \theta$. Update the average value of a point: $\theta \leftarrow \theta + \frac{1}{m}(v_c - v_d)$. Replace x_d in the population with x_c.

4. *(Reduce noise/Convergence test)* Set $t \leftarrow t + 1$. Set $p \leftarrow p_{max}\left(\frac{1}{t}\right)^r$. If $p \ge p_{min}$, go to step 2, otherwise go to step 1.

Parameters

Symbol	*Range*	*Value*	*Description*
m	$m > 0$	50	Size of population
p_{max}	$0 \le p_{max} \le 1$	0.5	Starting mutation probability
p_{min}	$0 \le p_{min} \le p_{max}$	0.035	Minimum mutation probability
r	$0 < r < 1$	$\frac{1}{4} + \frac{2}{n}$	Mutation reduction exponent

Figure 13-6: IGS-O: Iterated genetic search–Ordered combination

1. *(Restart)* Select a population of m points $x_1, \ldots, x_m$ at random and evaluate them. Compute the average value of a point: $\theta \leftarrow \frac{1}{m}\sum_{1\leq i\leq m} v_i$. Set $t \leftarrow 0$. Set $p \leftarrow p_{max}$.

2. *(Crossover/Mutate)* Pick two points x_i, x_j from the population at random. Select an integer k from 0 to n at random. Create a third point x_c as follows. Set $x_c \leftarrow \{b_1^i, \ldots, b_k^i, b_{k+1}^j, \ldots, b_n^j\}$. Then for each bit in x_c, with probability p set its value to *0* or *1* randomly. Evaluate x_c.

3. *(Reproduce)* Pick a point x_d from the population at random subject to the constraint that $v_d \leq \theta$. Update the average value of a point: $\theta \leftarrow \theta + \frac{1}{m}(v_c - v_d)$. Replace x_d in the population with x_c.

4. *(Reduce noise/Convergence test)* Set $t \leftarrow t+1$. Set $p \leftarrow p_{max}\left(\frac{1}{t}\right)^r$. If $p \geq p_{min}$, go to step 2, otherwise go to step 1.

Parameters

Symbol	*Range*	*Value*	*Description*
m	$m > 0$	50	Size of population
p_{max}	$0 \leq p_{max} \leq 1$	0.5	Starting mutation probability
p_{min}	$0 \leq p_{min} \leq p_{max}$	0.035	Minimum mutation probability
r	$0 < r < 1$	$\frac{1}{4} + \frac{2}{n}$	Mutation reduction exponent

Figure 13.5 presents the first of these algorithms, IGS-U. A population of m random points is created and evaluated. The probability of mutating each bit is set to its maximum value. Then, on each iteration, two points are selected at random and combined on a bit-by-bit basis. Additionally, each bit has a probability of being set to *0* or *1* at random. In step 3, a member of the population is deleted to make room for the new point. This is done by picking *uniformly* from the *below average* members of the population. In other words, the probability that an above-average member will be displaced is zero, and the probability that an average or below-average member will be displaced is $1/m_{bad}$, where m_{bad} is the number of members of the population with average or below average function values. In step 4, the probability of mutation is recomputed—getting monotonically smaller as t increases[2]—and finally the mutation rate is checked against a threshold to determine whether to continue or to restart the algorithm.

[2]Note that t does not actually represent time in terms of function evaluations, since it does not include the m evaluations required to initialize the population. I adopted this method of computing and reducing the mutation probability from the R1 algorithm (Holland, 1975).

The IGS-U algorithm implements basically the "survival of the fittest" algorithm just described, except that instead of picking losers in proportion to their badness, a boolean comparison with the average function value is used to determine eligibility for deletion. This implementation sacrifices some of the analytic results underlying genetic algorithms, but it has two significant advantages. The first advantage is generality. Genetic algorithms for function optimization have traditionally been restricted to positive function values, so that fitness can be computed by the ratio of a function value to the average value of the population. By using a boolean comparison instead of a division, the restriction to positive function values is unnecessary.

The second advantage is practical: The boolean approach *guarantees* the survival of an above-average point. In some cases a great deal of search may have gone into the discovery of an above-average point (and such a point may not be *way* above average—it depends on the function and the history of the search), but with the R1 approach such a point may have to survive many random $\frac{1}{m}$ chances of deletion before extra copies of the new point begin to appear in the population. With the relatively small populations employed in most genetic algorithms studies (usually $m < 100$), this can amount to a non-trivial source of loss.

13.2.6 Iterated genetic search–Ordered combination (IGS-O)

Readers familiar with genetic algorithms research will have noted a missing element in the definition of IGS-U—the combination rule (step 2 in Figure 13-5) does not employ the "cut-and-swap" crossover rule suggested by Holland (1975). Instead of picking a one (or perhaps two) crossover points, and combining ordered segments of the bit vector together, IGS-U simply makes a uniform random per-bit choice as to which "parent" should supply the needed bit value. The uniform choice rule embodied in IGS-U is easier to motivate in terms of spanning hamming subspaces (see Ackley 1987, Chapter 2, for discussion), but has generally been avoided by genetic algorithms researchers. The reasoning for this seems to have been that since uniform combination lacks an explicit coupling mechanism to hold co-adapted bit values together, it must be an inferior search technique.

To gather some empirical data on this question, I also tried an algorithm using a one-point cut-and-swap crossover rule. IGS-O is identical to IGS-U in definition, parameters, and parameter values, except that it employs the ordered combination rule rather than the uniform combination rule. The algorithm appears in Figure 13-6.

13.2.7 Stochastic iterated genetic hillclimbing (SIGH)

Use the SIGH algorithm as described in (Ackley 1987). Figure 13-7 summarizes the algorithm and the parameters. Note that the temperatures of the unit decisions, the amount of apathy for the voter unit decisions, and the payoff rate all depend on n. Note

Figure 13-7: SIGH: Stochastic iterated genetic hillclimbing

1. *(Initialize)* Create a group e of n position units and a group f of m voter units. Link each position unit to each voter unit with a symmetric bidirectional link with a variable link weight w_{ef}. Set all nm weights $w_{ef} \leftarrow 0$. Set $\theta \leftarrow 0$.

2. *(Election)* Apply the decision rule to the units of group e: For each $i \in e$, compute $\mathcal{I}_i = \sum_{j \in f} w_{ij} s_j$, and set s_i to $+1$ with probability $\frac{1}{1+e^{-\mathcal{I}_i/T_e}}$, and to -1 otherwise.

3. *(Reaction)* Apply the decision rule to the units of group f: For each $i \in f$, compute $\mathcal{I}_i = \sum_{j \in e} w_{ij} s_j$. With probability $1 - \frac{1}{e^{-(\mathcal{I}_i - \alpha)/T_f}}$, set s_i to $+1$, otherwise with probability $\frac{1}{e^{-(\mathcal{I}_i + \alpha)/T_f}}$, set s_i to -1, otherwise set s_i to 0.

4. *(Consequence)* Translate the state of group e to a bit vector x_n: $\boxed{+1}_e \Rightarrow 1$, $\boxed{-1}_e \Rightarrow 0$. Evaluate x_n. Let $r \leftarrow \frac{2}{1+e^{(v_n - \theta)/T_r}} - 1$. Add $\Delta w_{ij} = rks_i s_j$ to each weight w_{ij}. Compute $\theta \leftarrow \rho\theta + (1-\rho)(v_n + \delta)$. Go to step 2.

Parameters

Symbol	*Range*	*Value*	*Description*
m	$m > 0$	50	Size of population
T_e	$T_e > 0$	$10n$	Temperature of e-unit decisions
T_f	$T_f > 0$	$10n$	Temperature of f-unit decisions
α	$\alpha \geq 0$	$5n$	Apathy window of f-unit decisions
ρ	$0 \leq \rho \leq 1$	0.75	θ retention rate
k	$k > 0$	$\ln n$	Payoff rate
T_r	$T_r > 0$	2	Temperature of payoff saturation
δ	$\delta \geq 0$	4	θ offset

also that m—the size of the population—is held constant at 50, so the overall space complexity of the model, which is dominated by the number of links in the network, is $O(n)$.

13.3 Six functions

These demonstration functions were designed, essentially, by a kind of backwards analysis. We know how each search will end: the global maximum will be evaluated. Given that all of the algorithms tested are capable of some form of hillclimbing, the collecting area around the global maximum takes on obvious significance. In terms of search strategies, therefore, one way to get a low-dimensional visualization of the structure of a high-dimensional space is to consider what happens to the function value as one gets further from the global maximum. Do the values drop off smoothly all the way to the farthest reaches of hamming space? If so, simple hillclimbing will do the job. Do the values drop off for a while, then begin to rise again? If so, then local maxima must be considered, and more complex search strategies may be required. By constructing high-dimensional functions with specific function value vs. distance-from-max curves, a number of stereotypical properties of search spaces can be illustrated. The first four functions tested can be characterized in this fashion, since in those cases the function value depends only on the distance from the global maximum.

In this paper, the symbol c will be used to represent the number of matches between a given point and the global maximum. If a given point is the global maximum, then $c = n$, otherwise $c < n$. If a given point is the complement of the global maximum, then $c = 0$. ("c" was selected to suggest "hamming closeness," the complement of "hamming distance.")

This section presents each of the functions in the test suite, and shows the performances of the search strategies applied to each problem. Each test function was defined to embody a different stereotypical property of a search space, and in that sense the functions are very different from each other. However, there are two elements which are common to all of the functions, and these will be discussed first.

The first common element is that the global maximum is always located at the point designated by all *1*'s in the input vector. This means that the value of c for any given input vector is simply the number of *1* bits in the vector. Since none of the search strategies make any *a priori* distinction between *0* and *1*, this commonality does not imply any loss of generality in the test suite. Although some of the functions can be thought of as depending on the number of *1*'s, they could just as well be defined in terms of number of matches between an input vector and an arbitrary vector selected as the location of the global maximum. The all *1*'s corner of the hypercube is used to make the displays of the behavior of the algorithms easier to interpret.

The second common element is that the value of the global maximum is always $10n$, and with two minor exceptions, the minimum value of the space is always 0. This commonality *does* restrict the generality of the test suite, since some of the algorithms make assumptions about the range of possible function values. IHC-SA and IHC-NA do not depend on the actual range of the function values, since they only rely on boolean comparisons between function values to determine their behavior. The speed of these two algorithms is unaffected by a translation or (positive) scaling of the function space: given a function $f(\cdot)$, these algorithms search $af(\cdot) + b$ for positive a and real b at the same speed (on average) that they search f.

SHC, ISA, and SIGH rely on the numerical differences between function values to determine their behavior, and for that reason they can be affected by scale factors in the function value. However, they are insensitive to translation; they will search $f(\cdot) + b$ at the same speed that they search $f(\cdot)$.[3]

By contrast, a traditional genetic algorithm is sensitive to translation but not scaling, since the reproductive fitness of any given individual is determined by dividing its function value by the average of the space. As mentioned in Section 2.5, such algorithms assume positive function values and a meaningful zero point, so that *ratios* of function values can be interpreted as representing quantitative degrees of fitness. However, the two IGS algorithms do not actually perform this division, and fitness for reproduction is viewed qualitatively, based on a boolean comparison of the function value with the average value of the space (step 3 in Figures 13-5 and 13-6.) The basic IGS algorithms, therefore, are unaffected by translation or scaling. The IGS algorithm was chosen for testing, rather than a more conventional genetic algorithm such as R1 (Holland 1975), because during exploratory simulations it was found to perform much better than the R1 algorithm on every function in the test suite.[4]

13.3.1 A linear space—"One Max"

The first function in the test suite is a linear function, defined as follows: $f(x) = 10c$. The value of the function is simply ten times the number of *1* bits in the input vector x. This function presents no particular difficulty, since one hillclimb from anywhere in the space suffices to find the maximum; it is included in the test suite to investigate how well algorithms designed for more complicated situations perform in benign circumstances. Figure 13-8 tabulates the simulation results. (In all of these tables, the algorithms are listed in order of increasing time on the $n = 20$ problem.)

As expected, the two iterated hillclimbers found the maximum more quickly than did the other algorithms. IHC-NA is optimal in this application, on average evaluating less than n points before finding the maximum. IHC-SA took longer, since it considered

[3]This is not completely accurate in the case of SIGH, since translating the function space will affect the "startup transient" period when θ is moving from its initial zero value toward the sampled average value of the space. Limited simulation experiments varying b between ± 1000 suggest that this does not have much impact on the resulting search times.

[4]The interested reader is referred to (Ackley 1985) for some simulation results of R1 on a test suite that partially overlaps the present one.

One Max				
n	8	12	16	20
Algorithm	*Evaluations performed**			
IHC-NA	7	11	15	19
IHC-SA	30	70	126	209
SIGH	75	182	258	330
IGS-U	105	194	281	369
IGS-O	133	224	331	409
ISA	128	319	463	619
SHC	56	209	504	2,293

* Rounded averages over 50 runs.

Figure 13-8: Comparative simulation results on the "One Max" function. In all simulations, the performance measure is the number of objective function evaluations performed before the global maximum is evaluated.

changing bits that it had already considered changing before, but it still managed to come in second. The observed times for IHC-NA and IHC-SA agree well with their respective expected times of $n-1$ and $n^2/2$. SIGH and the IGS algorithms follow, taking less than twice as long as IHC-SA, and ISA and SHC bring up the rear.

A great deal of insight into the behavior of all of these algorithms can be gained by examining the sequence of points that they choose to evaluate on the way to finding the global maximum. In fact, if one takes seriously the idea of emergent properties and emergent levels of description, it is very important to take a hard look at the actual behavior of a system and try to understand what it is doing (even if, or perhaps especially if, it does not seem to be doing what it "should.") Figure 13-9 presents the median run of each of the search strategies on the One Max function. Each run is presented as a pair of synchronized graphs, the lower graph depicting function inputs and the upper graph depicting function values. In both cases, time, measured in function evaluations, runs horizontally. Note that the graphs of the different algorithms are *not* necessarily plotted on a common time scale—the behavior of each algorithm is shown at the finest time resolution (in powers of 2) that allows the entire search to fit onto the graph. At the beginning of the legend for each pair of graphs, the compression factor is displayed; for example, "[x1]" means that every function evaluation is plotted, and "[x4]" means that every fourth function evaluation is plotted.[5] Do not be misled by comparing lengths of

[5]There is one subtlety involved in the compression of the graphs. If it happens that the global maximum is evaluated on a trial that is not a multiple of the compression factor, that evaluation is "referred forward" to the next higher multiple of the compression factor, so that it will appear in the graph. For example, if the global maximum was evaluated on trial 3997 and the compression factor was [x4], the graph would

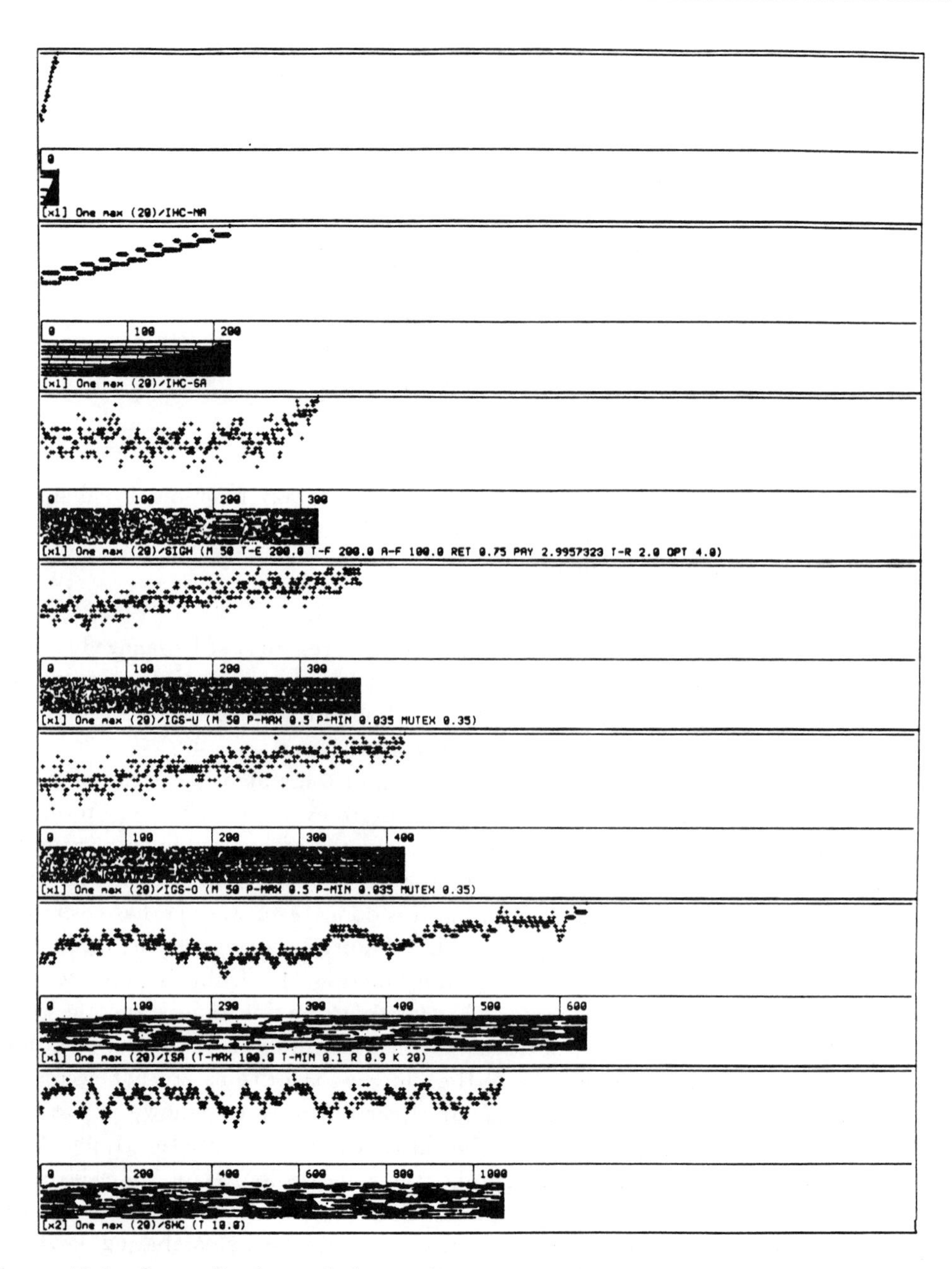

Figure 13-9: State displays of the median run of the seven search algorithms on the twenty bit "One Max" function.

graphs without checking the corresponding compression factors.

The upper graph of each pair displays function values vertically, with horizontal axes marked at the level of the global maximum and the global minimum. The lower graph displays the points evaluated by the search strategy at each time. Each binary vector is drawn vertically, with a dot representing a *1* bit and the absence of a dot representing a *0* bit. The global maximum, therefore, is represented by a completely black column. (The technology used to reproduce this document has limited resolution and accuracy, and it may sometimes be difficult to tell whether a given black spot on the page actually represents a *1* bit as opposed to being just "printing noise." I accept this level of error—rather than, say, using much bigger displays—because my main purpose in presenting these displays is not to supply exact down-to-the-bit data, but to show the broad characteristics of an algorithm, across an entire search, in a visually coherent form.)

The state graphs are presented on the page in the same order as the data is ranked in the corresponding table. The top graph in Figure 13-9 shows the median run of IHC-NA. The state graph shows an initially random pattern being overrun by a black wedge. The wedge arises from the increasing number of *1* bits in the points being evaluated; it expands upward from the bottom of the display because the algorithm enumerates the dimensions from bottom to top (as does IHC-SA and IGS-O). Although it may be difficult to discern for IHC-NA, the function values in the upper graph are not monotonically increasing—remember that the graphs are displaying the sequence of *points evaluated*, not the sequence of *current points*. If IHC-NA encounters a bit that is already a *1*, it tries changing it to a *0*, and on the One Max function, that produces a lower function value.

IHC-SA appears next, and the reason for its slowness compared to IHC-NA is obvious. Now instead of a single black triangle we see a series of diagonal lines, the result of evaluating all the neighbors of the current point. In the One Max function all uphill steps are the same size, and when there is more than one steepest ascent path IHC-SA takes the first one it found, so after each diagonal sweep the bottommost *0* bit is changed to a *1*, and another sweep commences.

The remaining algorithms do not display such geometric regularity, since randomness is incorporated into them more deeply than just in the selection of starting points. In all of these cases there is an initially chaotic period, and then gradually *1*'s predominate over *0*'s until eventually the global maximum is evaluated. In the case of IGS, the early chaotic period arises from the random initial population; in the case of ISA, it arises from the random pattern of moves made due to the initially high temperature; in SIGH it arises from the random behavior of the stochastic network starting from zero weights. SHC comes in at the bottom on One Max, because the fixed temperature of 10 causes it to head downhill fairly often. The graphs of ISA and SHC have a much more "streaky" appearance compared to SIGH and the IGS algorithms, due to the fact that

show the global maximum as occurring at trial 4000.

they consider only single bit changes. Successive points evaluated by ISA and SHC differ in no more than two bits.

Note that in the median run of SIGH, there are a few periods of semi-convergent search before the convergence that discovers the global maximum. The most obvious of these occurs roughly between trial 200 and trial 230; at that point it is searching a region several bits away from the global maximum. Although it is possible to hillclimb to the maximum from that convergence point (since One Max is a linear function), SIGH returns to more global search for a while before homing in on the optimal region of the space and finding the global maximum. Sometimes SIGH converges to a point one bit away from the global maximum, but abandons it before discovering the global maximum. With different parameter values (for example, a smaller δ), SIGH shows more persistence before giving up on a convergence. I tried such parameter values while tuning SIGH on the test suite, but it searched faster, on average, with the more "impatient" parameter values. On the One Max function, SIGH locates the correct region of the space easily, so the penalty for this behavior is not too severe.

13.3.2 A local maximum—"Two Max"

Linear functions are very easy to search. More complex spaces usually have local maxima, or "false peaks," that can lead a simple hillclimber away from the global maximum and eventually trap it. Here is a function with one global maximum and one local maximum: $f(x) = |18c - 8n|$. This function has the global maximum (with value $10n$) when the input is all *1*'s (so $c = n$), but it also has a local maximum (with value $8n$) when the input is all *0*'s (so $c = 0$). The boundary between the two maxima occurs at $\frac{4}{9}n$—if c is greater than that, hillclimbing will lead to the global maximum; if c is less, hillclimbing will lead to the local maximum.

Figure 13-10 tabulates the simulation results. With this function, a simple hillclimber may get stuck on the local maximum, so multiple starting points may be required. Nonetheless, on this function the relative speeds of the algorithms are identical to those on the One Max function. IHC-NA and IHC-SA still head the list. The mere existence of a local maximum does not imply that a space will be hard to search by iterated hillclimbing. The regions surrounding the two maxima of the function have a constant slope of 18 points per step toward the nearer maximum. The slopes have the same magnitude, so the higher peak must be wider at its base. With every random starting point, IHC-NA is odds on to start in the "collecting area" of the higher peak. Note that, *assuming* it is in the correct watershed, it is effectively searching a linear space, so it continues to excel.

A point to notice is that SIGH and SHC performed consistently *faster* on the Two Max function than they did on the One Max function. This is probably due to the steeper slope around the global maximum—18 points per bit for Two Max vs. 10 points per bit for One Max. Although the algorithms may spend a significant amount of time wandering around in the vicinity of the local maximum, when they get into the collecting

Two Max				
n	8	12	16	20
Algorithm	*Evaluations performed**			
IHC-NA	25	22	28	35
IHC-SA	37	91	176	271
SIGH	93	143	233	288
IGS-U	140	221	292	364
IGS-O	128	260	386	494
ISA	144	350	516	840
SHC	92	277	771	1,590

* Rounded averages over 50 runs.

Figure 13-10: Comparative simulation results on the "Two Max" function.

area of the global maximum, there is more pressure toward the maximum. In SIGH, the reinforcements tend to be larger. SHC is more likely to accept uphill moves, and reject downhill moves.

As in One Max, in Two Max IGS-U beats IGS-O by a significant margin. This is just as predicted, since neither function possesses structure in the ordering of the bits. On the median runs, IGS-O had more difficulty than IGS-U did with "lost alleles"—bit positions that, measured across the population, were mostly *0*, even though *1* was the better value. This is visible in the relatively long whitish streaks in the state display of IGS-O. In the run displayed, apparently a point with several of the topmost bits *1*—but a number of the lower bits *0*—was discovered fairly early on (around trial 150). Note that since IGS-O is sensitive to the ordering of the bits, a compact block of *1*'s such as this receives preferential treatment in the reproduction process. The spread of this point in the population apparently lead to a temporary reduction in the number of *1* bits in the lower bit positions of the pool, delaying somewhat the discovery of the global maximum.

Note how SIGH spends a significant period of time exploring the region around the local maximum (around trials 175–225) before suddenly shifting to the region of the global maximum. This behavior arises from the "distinction detector" characteristics of the units in the connectionist network—any changes to (say) the positive pole of a unit's receptive field cause identical effects (except bit-wise complemented) at the negative pole. Discovering that points with mostly *0*'s have good scores *automatically generates* the "hypothesis" in the network that points with mostly *1*'s will have good scores too. When the search diverges from the region around the local maximum, the region around the global maximum is the prime candidate for the next place to look.

A final point to note is that the median run of SHC on Two Max is actually faster

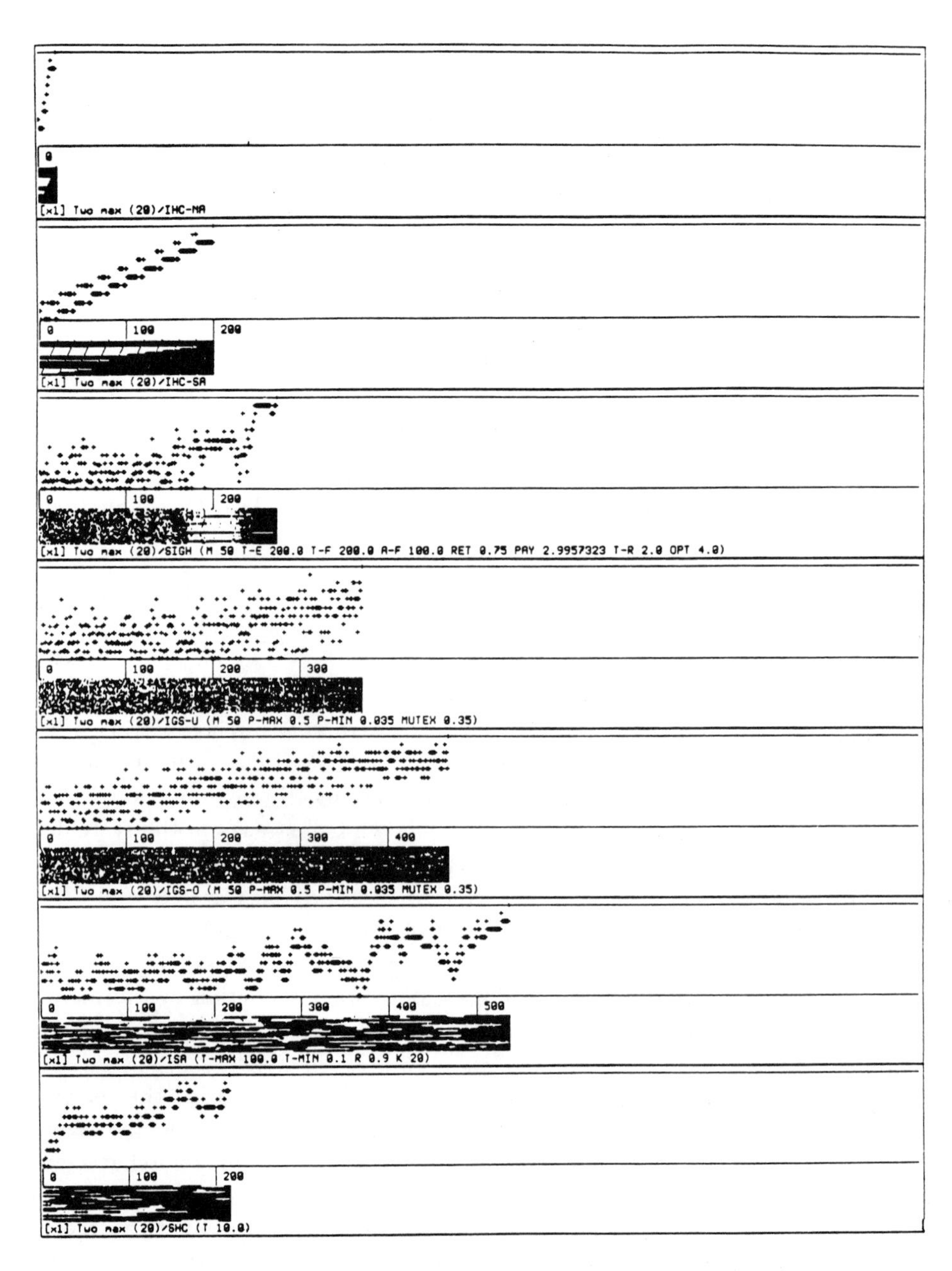

Figure 13-11: State displays of the median run of the seven search algorithms on the twenty bit "Two Max" function.

than the median run of all the other algorithms except the iterated hillclimbers. The explanation for this is simply that the runs of SHC form a largely bimodal distribution, depending on whether the algorithm does or does not get trapped in the collecting area of the local maximum. The median run of SHC is quite fast, but the occasional much longer runs due to entrapment pull the average up.

13.3.3 A large local maximum–"Trap"

The local maximum in the previous function presented little difficulty to any of the search strategies—its collecting area was small, so it was easy to avoid. What would happen, however, if the collecting area of a local maximum was much bigger than that of the global maximum? The "Trap" function, which is a variation on "Two Max," provides a test case.

In the Two Max function, the slopes surrounding the two maxima are equal in magnitude. In the Trap function, the slope is much steeper around the global maximum, and its collecting area is much smaller. Here is its definition:

Let $z = \left\lfloor \frac{3}{4}n \right\rfloor$. Then

$$f(x) = \begin{cases} \frac{8n}{z}(z - c) & \text{if } c \leq z, \\ \frac{10n}{(n-z)}(c - z) & \text{otherwise.} \end{cases}$$

As in the Two Max function, in the Trap function the global maximum equals $10n$ and occurs when all bits are one; the local maximum equals $8n$ and occurs when all bits are zero. The difference is that only the points that are less than $\frac{1}{4}n$ bits away from the global maximum are in its collecting area; hillclimbing from any other point leads to the local maximum. In the twenty bit space, only $\binom{20}{4} = 4{,}845$ states lead uniquely uphill to the global maximum. Another $\binom{20}{5} = 15{,}504$ states are global minima with value zero. In such states, every adjacent point is uphill, but three-quarters of those uphill moves lead away from the global maximum, toward the local maximum. In over 98% of the states in the space, all uphill moves lead away from the global maximum.

Figure 13-12 displays the results. On average, SIGH found the maximum of the twenty bit space in only 780 function evaluations. IHC-NA and IHC-SA ranked second and third, coming in at about four and nine times slower than SIGH, respectively. ISA averaged almost 200 times slower than SIGH, and somewhere in the fifty simulation runs, SHC, IGS-O, and IGS-U all exceeded 1,000,000 evaluations for a single search and were dropped out of the simulations. Note also that in the eight bit space, only SIGH found the maximum in appreciably less than the $2^8 = 256$ evaluations that random search would be expected to require. How could SIGH do so well?

Figure 13-13 suggests the answer. All of the algorithms, at first, hillclimb toward the local maximum. The iterated algorithms all get stuck there, and are restarted. When

Trap				
n	8	12	16	20
Algorithm	*Evaluations performed**			
SIGH	139	311	535	780
IHC-NA	234	743	2,039	3,522
IHC-SA	280	1,120	4,644	8,808
ISA	889	5,109	32,312	154,228
SHC	1,263	12,458	143,999	$> 1M$
IGS-O	532	17,158	$> 1M$	$> 1M$
IGS-U	459	26,703	$> 1M$	$> 1M$

* Rounded averages over 50 runs.

Figure 13-12: Comparative simulation results on the "Trap" function.

that happens, they are very likely to simply climb the local maximum again. One way to understand the relative success of IHC-NA and IHC-SA, compared to ISA and IGS, is that since they restart much more frequently than do the more sophisticated algorithms, they get more chances to stumble into the about 1.5% of the space that forms the collecting area of the global maximum. Like the other algorithms, SIGH initially hillclimbs toward the local maximum at all *0*'s. However, once it has converged at or near the local maximum, the effect of δ causes that state to be "taxed." As mentioned above with the Two Max function, and discussed in (Ackley 1987, Chapter 2), a voter unit is essentially a "distinction detector"—responding equally consistently to both the positive and negative poles of its receptive field. The initial convergence near the local maximum causes the receptive fields of the population to respond to *both* that state and its complement, even though the complementary states have not yet been sampled. Because of the design of the connectionist network used in SIGH—in particular, using ± 1 unit states and a product-of-states learning rule—*negative payoff for a state is equivalent to positive payoff for the complement of the state.* Consequently, when SIGH stops trying to climb up the local maximum, it does not return to completely random global search, the way that the iterated algorithms do. Instead, it is biased towards searching the complement of the local maximum. In the Trap function, the complement of the local maximum is the global maximum, and that is why SIGH performs so effectively.

Since the collecting area of the global maximum is so small, it typically takes several "tries" before SIGH gets close enough to sense the region around the global maximum, and this is clearly visible in the state graph. Note there are several brief periods of mostly *1* searching, around trials 250, 300, 475, and 550. In each of those cases, note the associated downward spike in the function values. SIGH didn't come close enough to the global maximum to show an improvement, and it "falls back" to the much

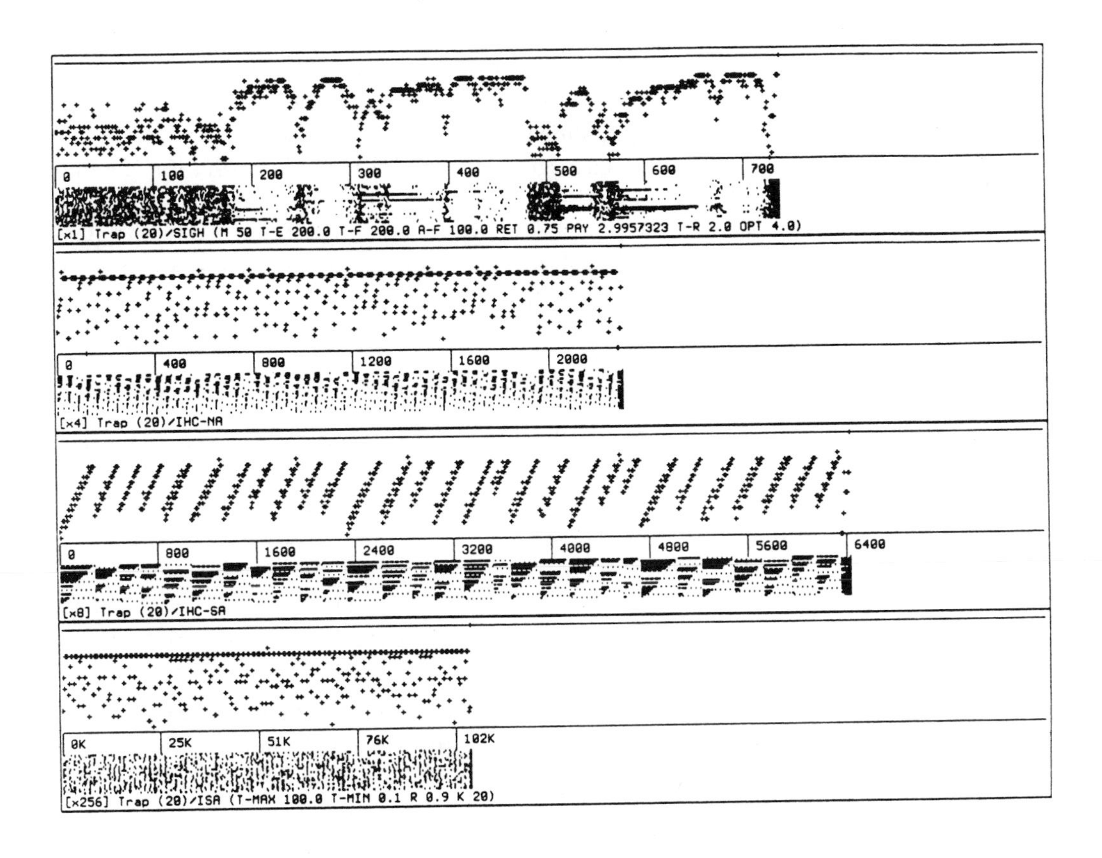

Figure 13-13: State displays of the median run of the four terminating algorithms on the twenty bit "Trap" function.

broader region around the local maximum. Finally, shortly after trial 700, it converges close to the global maximum, and optimizes the function shortly thereafter.

In effect, SIGH contains a heuristic that the other algorithms lack: *Try the opposites of good points*. For the Trap function, of course, this is ideal, but in general there is no reason to expect this to be advantageous.[6] Fortunately, SIGH is not *required* to spend much time searching the complement of a good point—it only will do so if it finds good function values there. For example, SIGH came in third on the One Max function, even though in that case the complement of a good point is a bad point. (And looking back at SIGH's behavior on One Max, shown in Figure 13-9, it is possible to see brief periods of largely *0* search following convergences—around trials 175–200, for example, and possibly even a little bit just before trial 100. Also, the behavior around trial 230 looks strongly like a "try the opposite" event. In that case, since the convergence point had several bits wrong, the complement had several bits right, producing a dip in the function value rather than a catastrophic swing.)

13.3.4 Fine-grained local maxima—"Porcupine"

Consider the following function: $f(x) = 10c - 15(1 - \text{parity}(c, n))$, where $\text{parity}(i, j)$ is 1 if i and j have the same parity—i.e., both odd or both even—and is zero if they have the different parity. In this test, n is always even, so we can describe this function as being just like the One Max function except that we subtract 15 when c is odd.[7] Every point that has an even number of *1* bits is a porcupine "quill," surrounded on all sides by the porcupine's "back"—lower valued points with odd numbers of *1* bits. As the total number of *1* bits grows, the back slopes upward; the task is to single out the quill extending above the highest point on the back.

Unlike the first two functions, the Porcupine function presents a tremendously rugged landscape when one is forced to navigate it by changing one bit at time. Not surprisingly, the iterated hillclimbers fail spectacularly here. Figure 13-14 displays the results. The landscape acts like flypaper, trapping IHC-NA after at most one move, and the resulting long simulation times reflect the exponential time needed to pick one of two points as a starting point: either the global maximum, or the point that differs from the global maximum only on the first dimension in the enumeration. From any other starting point, IHC-NA reaches a local maximum. IHC-SA does a little bit better than IHC-NA, because it will find the maximum if it starts at any point within one bit of the

[6] I also tried a function similar to Trap, except that I placed the local maximum at a location orthogonal to the global maximum (at a hamming distance of $n/2$), rather than at the complement of the global maximum. In that case, *all* of the "more sophisticated" algorithms (SHC, IGS-U, IGS-O, ISA, and SIGH) exceeded the time limit. Only the iterated hillclimbers succeeded, taking about as much time as they took on Trap.

[7] This function is one of the exceptions mentioned earlier—it does not have zero as the minimum possible value. Input vectors with exactly one *1* bit score -5. The other exception is the final function tested, which includes this function as a part, and so also has the potential to reach -5 as a minimum value.

Porcupine				
n	8	12	16	20
Algorithm	*Evaluations performed**			
SIGH	116	234	308	357
IGS-U	121	216	304	414
IGS-O	139	311	493	608
SHC	122	454	1,581	4,528
ISA	198	1,141	3,050	9,224
IHC-SA	335	6,633	101,143	$> 1M$
IHC-NA	1,211	26,652	$> 1M$	$> 1M$

* Rounded averages over 50 runs.

Figure 13-14: Comparative simulation results on the "Porcupine" function.

global maximum, so it has $n + 1$ chances to succeed rather than only two.

By contrast, the times required by SIGH, IGS-U, and IGS-O are less than a factor of two over their performances on the One Max function. The strong global property of the space—the more *1*'s the better, other things being equal—is detected and exploited effectively by SIGH and the IGS algorithms.

Figure 13-15 displays the median runs of the terminating algorithms. Note how SIGH spends a good deal of time at a local maximum (around trials 210–250) with two bits wrong. After that, it corrects one of the bits, even though that leads to a poorer function value. (It is difficult to see this in the function value graph, because the system is frequently trying *0*'s in some other position, which corrects the parity and leaves the function value unchanged.) At a local maximum all adjacent points are downhill, but in Porcupine, moves that increase the number of *1*'s are *less* downhill than those that decrease the number of *1*'s (-5 change in function value vs. -25). There is a brief destabilization (around trial 280) and then a convergence leading to the global maximum.

The Porcupine function turns out to be somewhat tricky for SHC and ISA. Both algorithms get down to two wrong or so, and eventually fix one of them, but then *every move* is uphill, so they are quite likely to simply move to a state with a different pair wrong. For example, notice the period from about trial 400 to about trial 800 in the ISA run. There are only a few pairs of *0*'s in the points being tested, and the algorithm shifts from one wrong pair to another several times. It eventually settles down on a state with four *0*'s. The system "freezes" at this point, and is restarted (around trial 1300). In the median run, ISA optimizes the function on its fifth try.

Although the Porcupine function reduced hillclimbing to random combinatoric search, in a sense it cheated to do so, by exploiting the hillclimber's extremely myopic view of possible places to move. A hillclimber that considered changing two bits at a

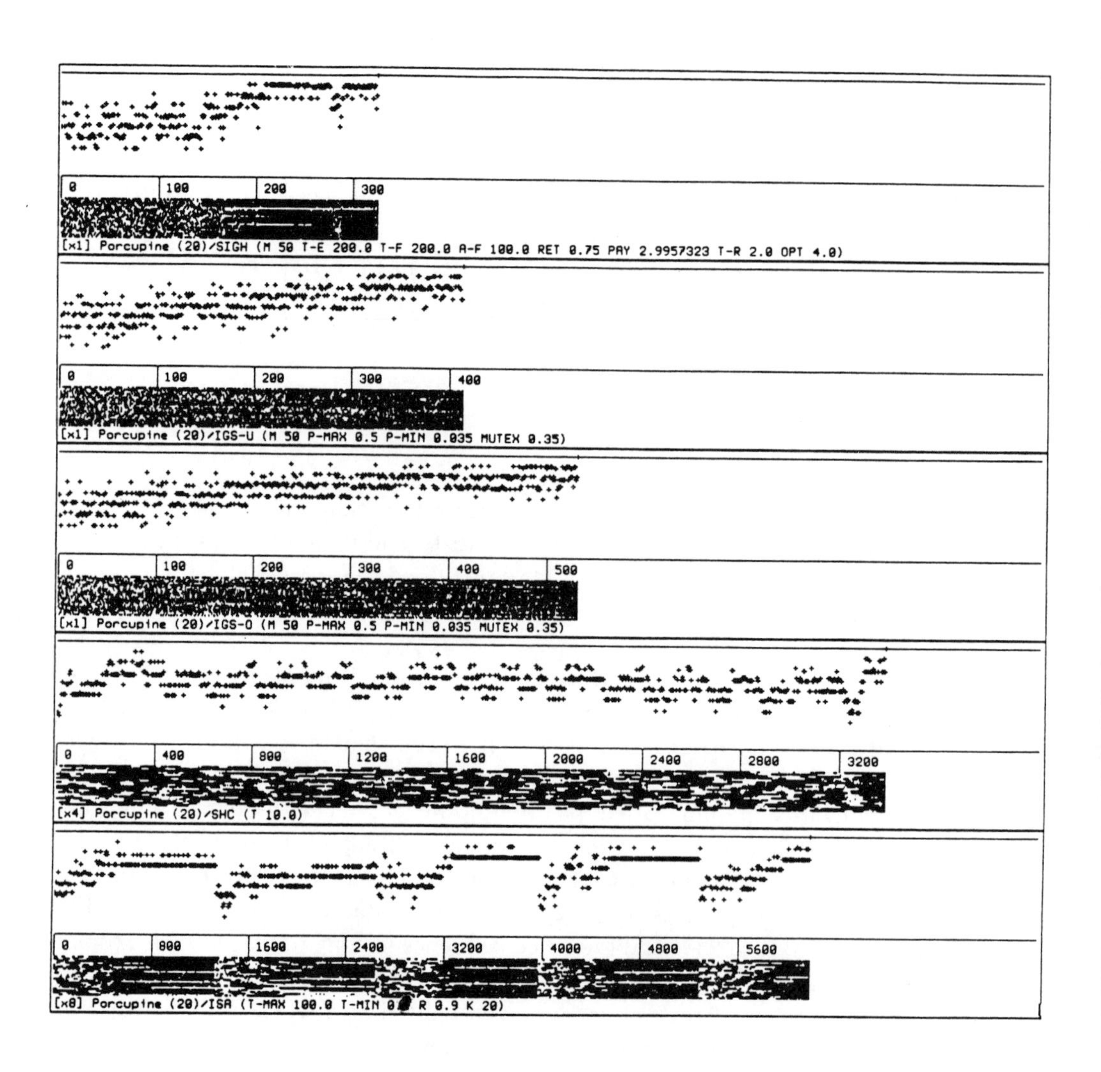

Figure 13-15: State displays of the median run of the five terminating search algorithms on the twenty bit "Porcupine" function.

Plateaus				
n	8	12	16	20
Algorithm	*Evaluations performed**			
SHC	40	105	227	494
IGS-U	111	217	389	779
ISA	111	268	481	814
IGS-O	119	230	371	910
SIGH	105	522	1256	2,979
IHC-NA	35	264	2,838	15,055
IHC-SA	71	398	2,340	23,433

* Rounded averages over 50 runs.

Figure 13-16: Comparative simulation results on the "Plateaus" function.

time could proceed directly to the highest quill. But increasing the working range of a hillclimber exacts its price in added function evaluations per move, and can be foiled anyway by using fewer, wider quills (e.g., subtract 25 points unless the number of ones is a multiple of three.) Higher peaks may always be just "over the horizon" of an algorithm that searches fixed distances outward from a single point.

13.3.5 Flat areas—"Plateaus"

The Porcupine function was full of local maxima, but they were all very small and narrow. A rather different sort of problem occurs when there are large regions of the space in which all points have the same value, offering no uphill direction. Consider the following "Plateaus" function: *Divide the bits into four equal-sized groups. For each group, if all the bits are 1 score* $2.5n$ *points, otherwise score* 0 *points. Return the sum of the scores for the four groups.* Unlike the other functions, this function has only five possible values: 0, $2.5n$, $5n$, $7.5n$, and $10n$. In a group, any pattern that has any *0*'s is on a plateau. Between the groups the bits are completely independent of each other; within a group only the *combined* states of all the units has any predictive power.[8]

Figure 13-16 displays the simulation results. SHC, which has been only a middling performer so far, is the winner in this case. Observing its behavior in Figure 13-17 shows why. When $n = 20$ there are five bits in each group. Any single bit flip will produce one of three outcomes—the function value will increase by 50 if the flip completes a group of *1*'s, or decrease by 50 if the flip disrupts a complete group of *1*'s, or remain the same if the flip does not change the all-*1*'s status of the group. Since SHC uses a temperature of

[8]Readers familiar with (Ackley 1985) should note that this is a slightly different plateaus function.

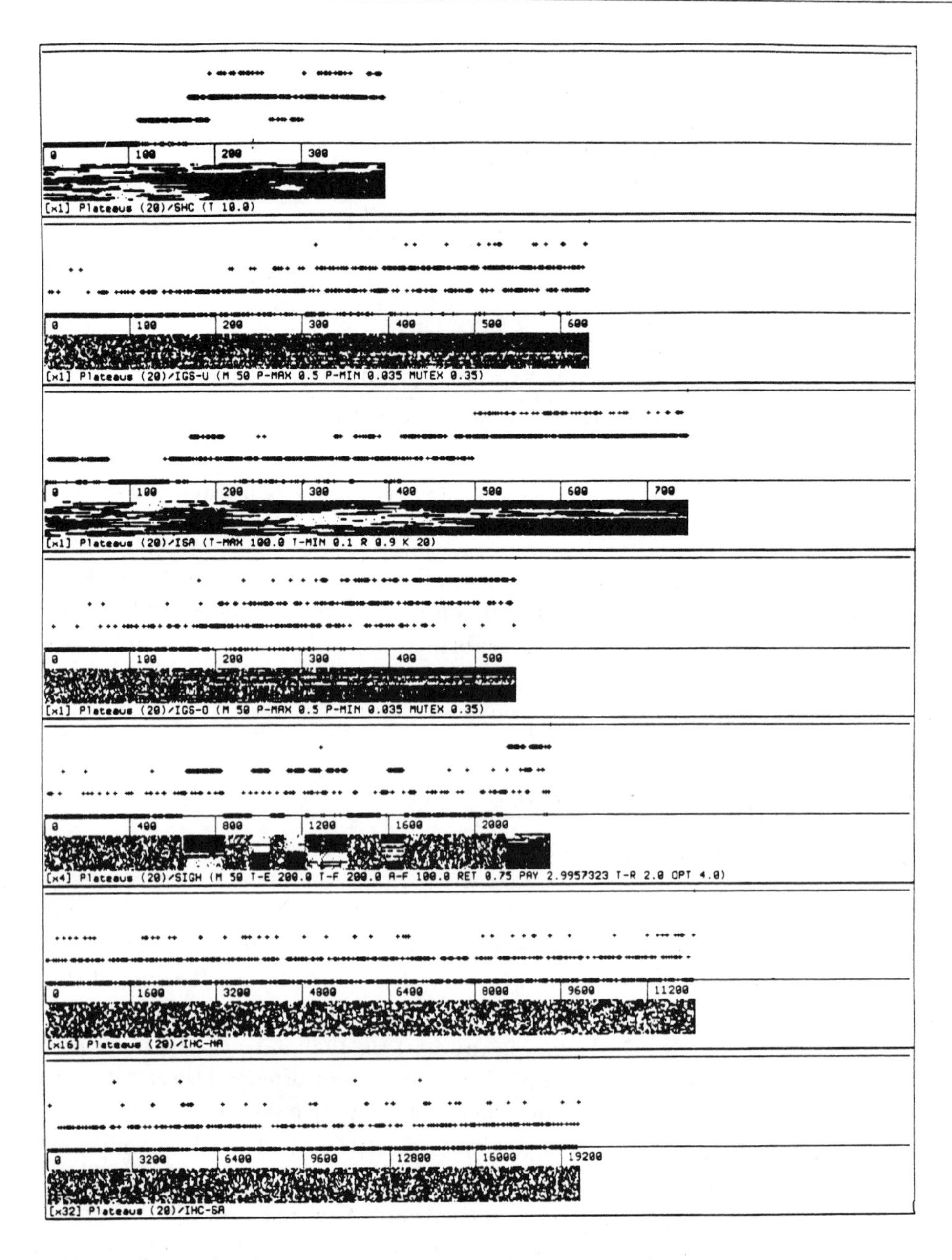

Figure 13-17: State displays of the median run of the seven search algorithms on the twenty bit "Plateaus" function.

10, swings of 50 in the function value means that when an improvement is found, SHC is very likely to accept it, and when a move causes a decrease, SHC is very likely to reject it. (In both cases, the probability is over .99). The typical behavior of SHC on this function, therefore, is to wander about the space randomly, and to "lock in" each group as the correct combinations of bits are encountered. This is the behavior evidenced in the median run of SHC, except for a period around trials 275–300, when it did accept a move that disrupted a completed group.

IGS-U comes in second on the Plateaus function. Note that any point that has i groups scoring 50 (in the $n = 20$ case) and the other $4 - i$ groups scoring 0 will produce a higher function value than every point with less than i groups correct. Once *one* such point is found, it is impossible for the population ever to contain only points with less than i groups correct. To see this, note that step 3 in Figures 13-5 and 13-6 ensures that above-average points are never candidates for deletion. Suppose there is exactly one point in the population with i groups maximized, and all other points have less than i groups maximized. In such a situation, the average value of the population must be less than the value of the point with i groups correct, so that point will be preserved. On the Plateaus function, the time-consuming part is finding the correct combinations, and SHC and the IGS algorithms, in effect, only have to find each combination once.

The Plateaus function is the first one considered in this suite in which the ordering of the bits can be relevant. In the twenty bit space, the value of each bit depends on four other bits. As can be discerned from the state displays, the related bits are adjacent in the bit vector representation. It is somewhat surprising, therefore, to note that IGS-O, which exploits that ordering information, still does on average somewhat worse than IGS-U, which does not. I do not have a satisfactory explanation for this effect at present, but note that the median run of IGS-O is actually somewhat faster than those of IGS-U and ISA, despite the difference in the averages.

Compared to its performance on the earlier functions, SIGH does significantly poorer on the Plateaus function, coming in near the bottom of the pack, beating only the iterated hillclimbers. The reason for this is that SIGH lacks the "lock in" characteristic of the genetic algorithms—SIGH, in general, requires multiple "hits" on a high-scoring point in order to cause a convergence. When the probability of getting a hit is so low, as in the Plateaus function, this need for multiple instantiations becomes expensive.[9] However, there seems to be an interesting story played out in the median run of SIGH on Plateaus—in the *way* in which the global maximum was located. Around trial 600, it converged on a point having the top two groups correct. (From the point of view of the search strategy, "*Aha!*"). Later on (around trial 900, and again around 1100), it converged on the complement of that point, which also had a relatively good score ("*As predicted!*") Up to this point, SIGH's behavior is reminiscent of its behavior on Two Max or Trap. However, in this case neither of those points achieves the success criterion, so the search continues. Then, shortly before trial 1600, SIGH converged on a third good

[9](Ackley 1987, Chapter 6) discusses this issue at greater length.

point, one that was almost orthogonal (at a distance of about $n/2$ in hamming space) to both of the previous good points (*"What?!"*) No such points exist in Two Max or Trap—in those cases, points orthogonal to the two maxima are not distinguished particularly, simply falling on the side of one hill or the other. This is followed by a relatively long period of global search (*"Hmm..."*), leading ultimately to a convergence at a point with three groups correct, and from there a stochastic hillclimb lead to the discovery of the global maximum (*"AHA!"*)

Across the runs of SIGH on Plateaus that I observed, the pattern of events that occurred in the median run of SIGH, while unique in detail, seemed typical in general: Converge on a good point, try the opposite, and if both are good try something different than either of them. Although at present I do not have a completely satisfactory explanation for why this general structure occurs, it seems to be a fairly robust phenomenon.

13.3.6 A combination space—"Mix"

Each of the functions considered so far has in some sense been a "pure essence"—a perfectly convex space, a space with exactly two maxima at opposite corners of the hypercube, and so forth. Such functions make for useful demonstrations of qualitative properties of search strategies, but they do not necessarily represent the sort of spaces that may arise in applications. In fact, it seems likely that "real functions" will not be so simple—they may well have a linear component, but will probably not be convex, they may well have paired maxima, but the maxima will not necessarily be all the way across the hypercube from each other, etc.

Consider this "Mix" function, which combines the properties of the other functions considered in this paper: *Divide the bits into five equal-sized groups. For the first group, score the bits according to the One Max function. For the second group, score the bits according to the Two Max function. For the third group, score the bits according to the Trap function. For the fourth group, score the bits according to the Porcupine function. For the fifth group, score the bits as though they were one group of the Plateaus function (i.e., score $2n$ points if the $n/5$ bits are all ones, and 0 otherwise). Return the sum of the four scores.* For this function, I upped the dimensionality of the space to 15–30 bits, because using smaller sizes made some of the "subfunctions" essentially degenerate. If $n = 10$, for example, each group is only two bits long, so it is not possible to be more than two bits away from the "global maximum" of each group. The range I tested provides from three to six bits per group, which allows a certain amount of complexity to be created.

Figure 13-18 displays the simulation results. Somewhat surprisingly (at least, I didn't expect it) SHC comes out on top. IGS-O, which so far has always been beaten by IGS-U, now takes a strong second. This agrees with expectations, since the bits in each of Mix's five little groups are adjacent to each other, so there is a strong structure

Mix				
n	15	20	25	30
Algorithm	*Evaluations performed**			
SHC	464	1,398	2,207	4,688
IGS-O	560	2,305	2,386	5,986
ISA	2,015	5,281	5,120	9,649
SIGH	1,937	3,981	6,342	12,477
IGS-U	762	3,664	4,146	17,374
IHC-SA	869	5,843	25,641	84,380
IHC-NA	1,059	11,973	14,362	148,494

* Rounded averages over 50 runs.

Figure 13-18: Comparative simulation results on the "Mix" function.

in the ordering of the bits.[10] ISA and SIGH follow, taking about half again as long and twice as long, respectively, as IGS-O. IGS-U is a little farther back, and the two iterated hillclimbers are much farther back.

Figure 13-19 displays the median runs. The groups are laid out in the state graphs from bottom to top, five bits each. At the bottom is the group for One Max, followed upwards by the groups for Two Max, Trap, and Porcupine, and the group for Plateaus is at the top. Different algorithms ran into different difficulties with Mix—in general, matching their strengths and weaknesses on the "pure functions." SHC exploited its ability to "latch onto" large changes in function value to retain the first discovery of the maximal point for the Plateaus group. In this particular run, the Porcupine and One Max groups were the last ones that SHC got right. IGS-O got trapped by Trap in its first run, but on its second try it homed in rapidly on the global maximum. ISA was caught twice by Trap, and once by Trap and Two Max, before optimizing the function.

Initially, SIGH was stopped by the Plateaus group—within the first 800 trials, it came within two bits of the global maximum, with both errors occuring in the Plateaus group. It was able to correct the Plateaus group fairly quickly, but when it did so it was in the local maxima for Two Max and Trap (around trials 1000–1500). This caused it quite a bit of trouble because it then tended to "associate" *1* bits in the Plateaus group with *0* bits in the Two Max and Trap groups. As the search continued, SIGH tried a number of variations such as getting Trap but not Two Max (around trials 2600–2800), or Two Max but not Trap (around trial 5400), and so forth. It returned to the region of its initial convergence (around trial 3200) but did not find the maximum. In this run,

[10] I could not resist trying the thirty-bit Mix simulations again and scrambling the bits randomly before the commencement of each search. IGS-O's average time rose to 15,850, placing it behind SIGH but still a bit ahead of IGS-U. The average times for SHC, ISA, SIGH, and IGS-U changed by less than 1%.

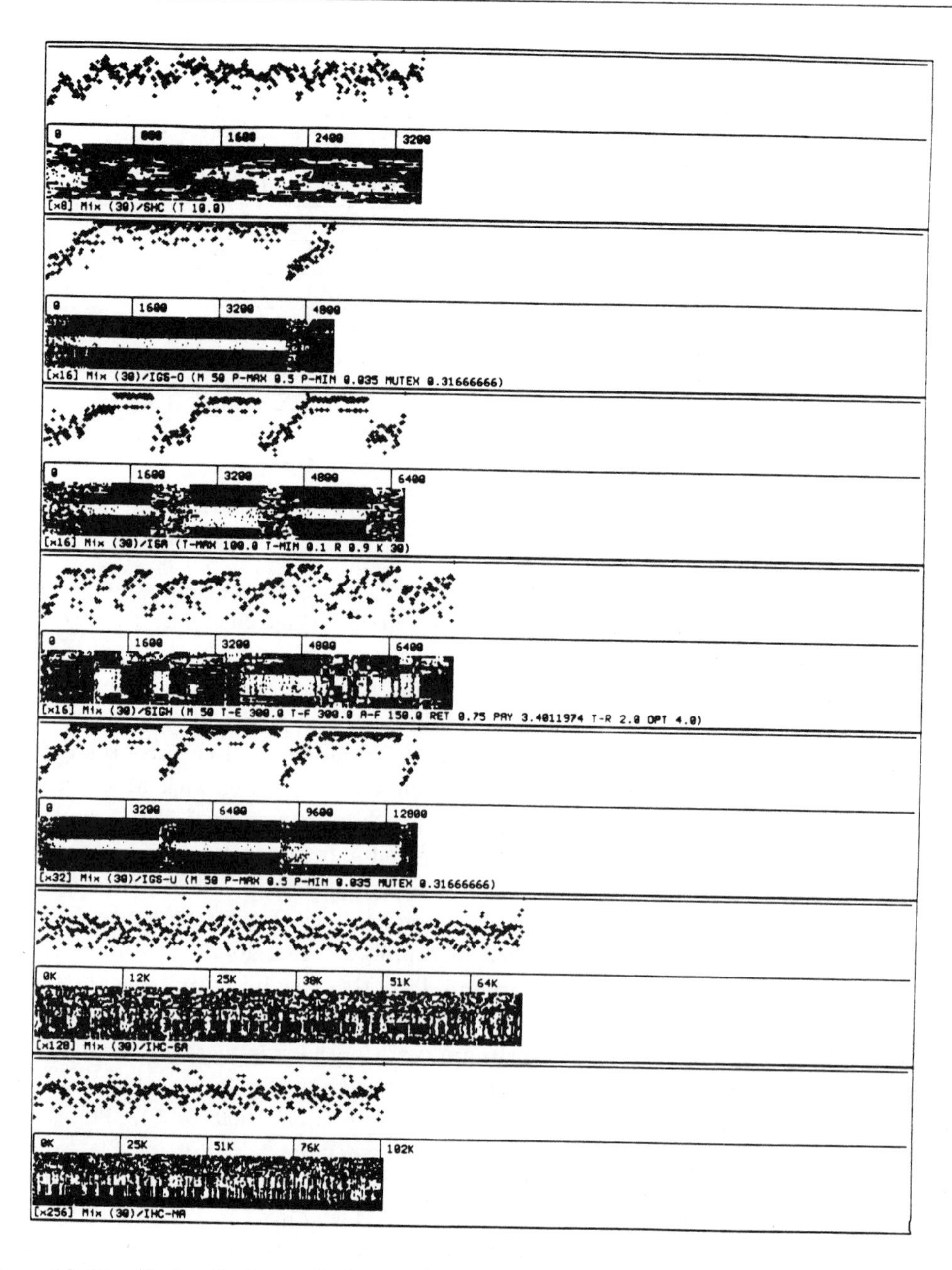

Figure 13-19: State displays of the median run of the seven search algorithms on the thirty bit "Mix" function.

the third time was the charm—SIGH returned once again to the region of the initial convergence (after trial 7000 or so) and finally found the global maximum.

The behavior of IGS-U looks qualitatively like IGS-O, the only difference being that it gets trapped by Two Max or Trap more frequently. In the median run it took three restarts to find the global maximum. The iterated hillclimbers, which did so excellently on the simple One Max and Two Max functions, once again find themselves at the bottom of the pile.

13.4 Conclusions

Given a set of results such as these, the urge to fold them down somehow into a single overall ranking is almost irresistible, and yet there is no obviously meaningful way to do that. It requires some means of assigning relative worth to the individual functions in the test suite—some reason to make claims such as "a good performance on Mix is ten times more valuable than a good performance on One Max"—but there seem to be no *a priori* grounds for making that assessment. There is simply *no* "best strategy" for solving all possible problems. In any real application, a wide range of factors beyond simply number of function evaluations must be considered. Is the strategy easy to implement on available hardware? Do parameters have to be tuned? Is the cost of evaluating the function at an arbitrary point really a constant?

Lacking answers to questions like these, I will not try to make any claim about "who's #1". I do make the following observations

1. None of the tested strategies is best for all of the tested functions.
2. In the tested functions that imposed no structure on the ordering of the bit vector, genetic search using uniform combination was faster than genetic search using standard crossover.
3. To the level of generality of the test suite, stochastic iterated genetic hillclimbing is competitive in speed with a variety of existing algorithms.

The black box function optimization problem, as I have formulated it, embraces all possible functions from a bit vector to the reals. No completely "general purpose" search strategy—capable of *quickly* optimizing every function—can exist. All search strategies make assumptions about the functions they will face, and that those assumptions will be wrong for some possible function. No search strategy is a panacea.

I defined the goal of the computation to be the satisfaction of a strictly defined success criterion. Until a point satisfying the criterion is found, the computation remains unfinished. This strong constraint satisfaction approach creates a situation in which permanent convergence—such as produced by an unaltered genetic algorithm or an unaltered simulated annealing algorithm—engenders the risk of an infinite loop: They may converge in the wrong spot. On the other hand, the exponential size of the search spaces rules out brute force enumeration or random search. To make progress in such circumstances, it is important to concentrate the search in the most promising region of the space, but it is just as important, later on, to back off of any one position, and—with lessons learned—to try again.

References

Ackley 1985 Ackley, D.H. A Connectionist Genetic Algorithm. In J. Grefenstette (ed.) *Proceedings of an International Conference on Genetic Algorithms and Their Applications*, 121–135. Carnegie Mellon University, Pittsburgh, PA.

Ackley 1987 Ackley, D.H. *Stochastic Iterated Genetic Hillclimbing*. Doctoral dissertation in computer science. Carnegie Mellon University. Pittsburgh, PA.

Ackley *et al* 1985 Ackley, D.H., Hinton, G.E., and Sejnowski, T.J. A Learning Algorithm for Boltzmann Machines, *Cognitive Science*, **9.1**, 147–169.

Albus 1979 Albus, J.S. Mechanisms of Planning and Problem Solving in the Brain. *Mathematical Biosciences*, **45**, 247–293.

Anderson 1983 Anderson, J. R. *The Architecture of Cognition*. Cambridge: Harvard University Press.

Axelrod 1980a Axelrod, Robert. Effective Choice in the Prisoner's Dilemma, *Journal of Conflict Resolution* **24**, 3-25.

Axelrod 1980b Axelrod, Robert. More Effective Choice in the Prisoner's Dilemma, *Journal of Conflict Resolution* **24**, 379-403.

Axelrod 1984 Axelrod, Robert. *The Evolution of Cooperation*. New York: Basic Books.

Axelrod and Hamilton 1981 Axelrod, Robert and William D. Hamilton. The Evolution of Cooperation, *Science* **211** 1390-1396.

Baker 1985 Baker, J.E. Adaptive Selection Methods for Genetic Algorithms. Grefenstette 1985, 101-111.

Ballard 1986 Ballard, D.H. Parallel Logical Inference and Energy Minimization. *Proceedings of the National Conference on Artificial Intelligence*, Philadelphia, August 1986, 203-208.

Barto 1985 Barto, A. Learning by Statistical Cooperation of Self-Interested Neuron-like Computing Elements. COIN Technical Report 85-11, University of Massachusetts.

Batcher 1968 Batcher, K.E. Sorting Networks and their Applications. In *Proceedings of the 1968 Spring Joint Computer Conference*, AFIPS, 307-314.

Becker 1985 Becker, J.D. Typing Chinese, Japanese, and Korean. *IEEE Computer*, **18**, no.1, 27-34.

Bethke 1981 Bethke, A.D. *Genetic Algorithms as Function Optimizers*. Ph.D. dissertation, University of Michigan, Ann Arbor.

Bonomi 1984 E. Bonomi and J-l. Lutton. The N-city Travelling Salesman Problem: Statistical Mechanics and the Metropolis Algorithm. *Siam Review*, **26**, No. 4, pp. 551-568.

Booker 1982 Booker, L.B. *Intelligent Behavior as an Adaptation to the Task Environment*. Ph.D. dissertation, University of Michigan, Ann Arbor.

Boorman *et al* **1980** Boorman, Scott A. and Paul R. Levitt. *The Genetics of Altruism*. New York: Academic Press.

Brindle 1981 Brindle, A. *Genetic Algorithms for Function Optimization*. Ph.D. dissertation, University of Alberta, Edmonton, Canada.

Brownston *et al* **1985** Brownston, L., Farrel, R., Kant, E., & Martin, M. *Programming Expert Systems in OPS5*. New York: Addison-Wesley.

Buchanan *et al* **1978** Buchanan, B.G., E.A.Feigenbaum. DENDRAL and MetaDENDRAL: Their Application Dimension. *Artificial Intelligence: An International Journal* **11**, 5-24.

Coombs and Davis 1987 Coombs, Susan and Lawrence Davis. Genetic Algorithms and Communication Link Speed Design: Implementation Issues and System Performance. To appear in Grefenstette 1987.

Cottrell 1985 Cottrell, G.W. *A connectionist approach to word sense disambiguation*. Doctoral dissertation, Computer Science Department, University of Rochester, Rochester, NY 14627, April 1985.

Cottrell and Small 1983 Cottrell, G.W. and Small, S.L. A connectionist scheme for modelling word sense disambiguation. *Cognition and Brain Theory*, **6**(1), 1983, 89-120.

Cui 1985 Cui, W. Evaluation of Chinese character keyboards. *IEEE Computer*, **18**, No.1, 54-59.

Davis 1985a Davis, L. Job Shop Scheduling with Genetic Algorithms. In Grefenstette 1985, 136-140.

Davis 1985b Davis, L. Applying adaptive algorithms to epistatic domains. In *Proceedings of the Ninth International Joint Conference on Artificial Intelligence (IJCAI 85)*, Los Angeles, **9**, 162-164.

Davis and Coombs 1987a Davis, L. and Susan Coombs. Optimizing Network Link Sizes with Genetic Algorithms. To appear in *Modelling and Simulation Methodology: Knowledge Systems Paradigms*, Maurice S. Elzas, Tuncer I. Oren, and Bernard P. Zeigler editors, North Holland Publishing Co.

Davis and Coombs 1987b Davis, L. and Susan Coombs. Genetic Algorithms and Communication Link Speed Design: Theoretical Considerations. To appear in Grefenstette 1987.

Davis and Ritter 1987 Davis, L. and Frank Ritter. Schedule Optimization with Probabilistic Search. *Proceedings of the 3rd IEEE Conference on Artificial Intelligence Applications.*

Davis and King 1977 Davis, R. & King, J. An overview of production systems. In E. W. Elcock and D. Michie (eds.) *Machine Intelligence* **8**. New York: John Wiley and Sons.

DeJong 1975 DeJong, K.A. *Analysis of the Behavior of a Class of Genetic Adaptive Systems.* Ph.D. dissertation, Dept. Computer and Communication Sciences, University of Michigan, Ann Arbor.

DeJong 1980 DeJong, K.A. Adaptive System Design: a Genetic Approach. *IEEE Transactions on Systems, Man, and Cybernetics*, **Vol.SMC-10**, No.9, September.

DeJong and Smith 1981 DeJong, K.A., and T. Smith. Genetic Algorithms Applied to Information Driven Models of US Migration Patterns. *12th Annual Pittsburgh Conference on Modeling and Simulation*, April.

Derthick 1986 Derthick, M.. *A Connectionist Knowledge Representation System.* Ph.D. thesis proposal, Department of Computer Science, Carnegie-Mellon University, Pittsburgh PA, June.

Eastman 1981 Eastman, C. Recent Developments in Representation in the Science of Design. In *Proceedings of the 18th Design Automation Conference*, IEEE Computer Society and Association for Computing Machinery.

Englander 1985 Englander, A.C. Machine Learning of Visual Recognition Using Genetic Algorithms. In Grefenstette 1985, 197-201.

Fahlman *et al* 1983 Fahlman, S.E., Hinton, G.E. and Sejnowski, T.J. Massively parallel architectures for AI: NETL, Thistle, and Boltzmann machines. *Proceedings of the National Conference on Artificial Intelligence*, Washington, August, 109-113.

Fanty 1985 Fanty, M. Context-Free Parsing in Connectionist Networks. Technical Report 174, Computer Science Department, The University of Rochester, Rochester NY, November.

Feldman 1982 Feldman, J.A. Dynamic Connections in Neural Networks. Biological Cybernetics **46**, 27-39.

Feldman 1985 Feldman, J.A. Energy and the Behavior of Connectionist Models. Technical Report 155, Computer Science Department, The University of Rochester, Rochester NY, November.

Feldman and Ballard 1982 Feldman, J.A. and Ballard, D.H. Connectionist models and their properties. *Cognitive Science* **6**, 1982, 205-254.

Fenves and Norabhoompipat 1978 Fenves, S.J. and T.Norabhoompipat). Potentials for Artificial Intelligence Applications in Structural Engineering Design and Detailing. In J.C.Latombe, ed., *Artificial Intelligence and Pattern Recognition in Computer Aided Design*

Fitzpatrick *et al* 1984 Fitzpatrick, J.M., J.J.Grefenstette, and D. Van Gucht. Image Registration by Genetic Search. In *Proceedings of IEEE SOUTHEASTCON '84*, Louisville, KY, April, 460-464.

Ford and Norton 1985 Ford, R. and J. Norton. An Expert System for Selection of Module Implementations. *Proceedings of the 1985 Conference on Intelligent Systems and Machines*, Rochester, MI, April.

Forgy 1981 Forgy, C.L. *OPS5 Manual.* Carnegie-Mellon University Computer Science Department Technical Report.

Forrest 1985 Forrest, S. Implementing Semantic Network Structures Using Classifier Systems. In Grefenstette 1985.

Fourman 1985 Fourman, M.P. Compaction of Symbolic Layout Using Genetic Algorithms. In Grefenstette 1985, 141-153.

Freeman and Newell 1971 Freeman, P., A.Newell. A Model for Functional Reasoning in Design. *Proceedings of the International Joint Conference on Artificial Intelligence*, **2**, 621-640.

Garey and Johnson 1979 M. R. Garey and D. S. Johnson. *Computers and Intractability.* W. H. Freeman Co., San Francisco.

Geman 1984 Geman, S. and Geman D. Stochastic Relaxation, Gibbs Distributions, and the Bayesian Restoration of Images. *IEEE Transactions on Pattern Analysis and Machine Intelligence* **PAMI-6**, 721-741.

Gidas 1985 B. Gidas. Non-stationary Markov Chains and Convergence of the Annealing Algorithm. In *Journal of Statistical Physics*, **39**, 73-131.

Glover 1986 Glover, D.E. Experimentation with an Adaptive Search Strategy for Solving a Keyboard Design/Configuration Problem. Computer Science Department Technical Report 86-03, University of Iowa, Iowa City, Iowa, May.

Goldberg 1983 Goldberg, D. E. *Computer-aided Gas Pipeline Operation Using Genetic Algorithms and Machine Learning.* Ph.D. Dissertation in Civil Engineering, University of Michigan.

Goldberg 1985a Goldberg, D.E. Dynamic System Control Using Rule Learning and Genetic Algorithms. In *Proceedings of the International Joint Conference on Artificial Intelligence*, *9*, 588-592.

Goldberg 1985b Goldberg, D.E. Genetic Algorithms and Rule Learning in Dynamic System Control. In Grefenstette 1985, 8-15.

Goldberg and Lingle 1985 Goldberg, D.E., R.Lingle, Jr. Alleles, Loci, and the Traveling Salesman Problem. In Grefenstette 1985 154-159.

Goldberg and Thomas 1986 Goldberg, David E. and Amanda L. Thomas. Genetic Algorithms: A Bibliography 1962-1986. TCGA Report No. 86001, The Clearinghouse for Genetic Algorithms, University of Alabama.

Golden 85 B. L. Golden and W. R. Stewart. Empirical Analysis of Heuristics. In *The Traveling Salesman Problem*, ed. E. L. Lawler, J.K. Lenstra, A. H. G. Rinnooy Kan, and D. B. Shmoys, John Wiley and Sons Ltd., 207-249.

Grefenstette 1984a John J. Grefenstette. GENESIS: A System for Using Genetic Search Procedures. In *Proceedings of a Conference on Intelligent Systems and Machines*, Rochester, MI, April, 161-165.

Grefenstette 1984b John J. Grefenstette. A User's Guide to GENESIS. report of Vanderbilt University.

Grefenstette 1985 Grefenstette, John, J. (ed.). *Proceedings of an International Conference on Genetic Algorithms and Their Applications.* Pittsburgh, PA: The Robotics Institute of Carnegie-Mellon University.

Grefenstette 1986 Grefenstette, J.J. Optimization of Control Parameters for Genetic Algorithms. In *IEEE Transactions on Systems, Man and Cybernetics*, SMC-16(1), 122-128.

Grefenstette *et al* **1985** Grefenstette, J.J., R.Gopal, B.Rosmaita, and D. Van Gucht. Genetic Algorithms for the Traveling Salesman Problem. In Grefenstette 1985, 160-168.

Grefenstette 1987 Grefenstette, J.J. ed. *Proceedings of the Second Conference on Genetic Algorithms and their Applications.*

Grinberg 1980 Grinberg, M.R. A Knowledge Based Design System for Digital Electronics. *American Association for Artificial Intelligence* **1**, 283-285.

Grossberg 1978 Grossberg, S. Competition, Decision, Consensus, Journal of Mathematical Analysis and Applications. **66**, 470-93.

Gupta 1984 Gupta, A. Parallelism in Production Systems: The Sources and the Expected Speed-up. Carnegie-Mellon University Computer Science Department Technical Report CMU-CS-84-169, December.

Hajek 1985 B. Hajek. Cooling Schedules for Optimal Annealing. Preprint, Department of Electrical Engineering, UNiversity of Illinois/Champaign-Urbana.

Hamilton 1980 Hamilton, William D. Sex versus Non-Sex versus Parasite. *Oikos* **35** 282-290.

Hamilton 1982 Hamilton, William D. Heritable True Fitness and Bright Birds: A Role for Parasites. *Science* **218** 384-387.

Harbison and Steele 1984 Harbison, S.P., and Steele, G.L. *C: A Reference Manual.* Prentice-Hall, New Jersey.

Hebb 1949 Hebb, D.O. *The Organization of Behavior.* New York: Wiley.

Hillis 1985 Hillis, W.D. *The Connection Machine.* MIT Press, 1985.

Hillis and Steele 1986 Hillis, W.D. and Steele, G.L. Data Parallel Algorithms. CACM **29:12** December, 1170-1183.

Hinton 1981 Hinton, G.E. Implementing Semantic Networks in Parallel Hardware. In G. Hinton and J. Anderson (Eds.), *Parallel Models of Associative Memory*, Hillsdale, NJ, Erlbaum, 161-187.

Hinton *et al* **1986** Hinton, G. E., McClelland, J. M. & Rumelhart, D. E. Distributed representations. In D. E. Rumelhart and J. L. McClelland (Eds.), Parallel Distributed Processing: Explorations in the Microstructure of Cognition, Volume 1. Cambridge, MA: Bradford Books.

Hinton and Sejnowski 1983 Hinton, G.E. and Sejnowski, T.J. Analyzing Cooperative Computation. *Proceedings of the Fifth Annual Conference of the Cognitive Science Society*, Rochester, NY, May.

Hinton and Sejnowski 1986 Learning and Relearning in Boltzmann Machines. In *Parallel Distributed Processing* by D.E. Rumelhart, J.L. McClelland, and the PDP Research group, vol. 1, Cambridge: Bradford/MIT Press, 282-317.

Holland 1968 Holland, John H. Hierarchical Descriptions of Universal Spaces and Adaptive Systems. (Technical Report ORA Projects 01252 and 08226), Ann Arbor: University of Michigan, Department of Computer and Communication Sciences.

Holland 1975 Holland, John H. *Adaptation in Natural and Artificial Systems.* Ann Arbor: University of Michigan Press.

Holland 1980 Holland, John H. Adaptive Algorithms for Discovering and Using General Patterns in Growing Knowledge Bases. *International Journal of Policy Analysis and Information Systems* **4** 245-268.

Holland 1985 Holland, J.H. Properties of the Bucket Brigade Algorithm. Grefenstette 1985, 1-7.

Holland 1986 Holland, J.H. Escaping Brittleness: The Possibilities of General Purpose Learning Algorithms Applied to Parallel Rule-based Systems. In Michalski, Carbonell, and Mitchell (Eds.), *Machine Learning II*, Morgan Kaufmann, Los Altos, CA.

Holland and Reitman 1978 Holland, J.H., J.R.Reitman. Cognitive Systems Based on Adaptive Algorithms. In D.A.Waterman, F.Hayes-Roth (Eds.), *Pattern-directed Inference Systems*, New York: Academic Press.

Holland, *et al* 1986 Holland, J.H., Holyoak, K.J., Nisbett, R.E., and Thagard, P.R. *Induction: Processes of Inference, Learning, and Discovery.* MIT Press.

Hopfield 1982 Hopfield, J. J. Neural Networks and Physical Systems with Emergent Collective Computational Abilities. *Proceedings of the National Academy of Sciences USA*, **79** 2554-2558.

Hopfield and Tank 1985 Hopfield, J.J. and Tank, D.W. "Neural" Computation of Decisions in Optimization Problems. *Biological Cybernetics* **52**, 141-152.

Huang 1985 Huang, J.K. The Input and Output of Chinese and Japanese Characters. *IEEE Computer*, **18**, No.1, 18-26.

Kim and MdDermott 1983 Kim, J., J.McDermott. TALIB: an IC layout design assistant. *Proceedings of the National Conference on Artificial Intelligence* 3, 197-201.

Kirkpatrick *et al* **1983** Kirkpatrick, S. Gelatt, C. D. & Vecchi, M. P. Optimization by Simulated Annealing. *Science*, **220** 671-680.

Kirkpatrick 1984 S. kirkpatrick. Optimization by Simulated Annealing: Quantitative Studies. *Journal of Statistical Physics*, **34**, Nos. 5/6, 975-986.

Klopf 1982 Klopf, A.H. *The Hedonistic Neuron.* Washington, DC: Hemisphere.

Laird 1985 Laird, J. *SOAR User's Manual.* Xerox PARC Internal Document.

Laird *et al* **1985** Laird, J., Rosenbloom, P., and Newell, A. Towards Chunking as a General Learning Mechanism. In *Two SOAR Studies*, Carnegie-Mellon University Computer Science Department Technical Report CMU-CS-85-110.

Laird *et al* **1986** Laird, J.E., Rosenbloom, P.S., and Newell, A. Chunking in Soar: The Anatomy of a General Learning Mechanism. *Machine Learning*, **1**, No. 1.

Lasser 1986 Lasser, C. The Essential *Lisp Manual. Thinking Machines Technical Report, July.

Lin 73 S. Lin and B. W. Kernighan. An Effective Heuristic Algorithm for the Traveling Salesman Problem. *Operations Research* **11**, 498-516.

Marsaglia 1968 Marsaglia, G. Random Numbers Fall Mainly in the Planes. *Proceedings of the National Academy of Sciences*, **61** 1, 25-28.

Mauldin 1984 Mauldin, M.L. Maintaining Diversity in Genetic Search. *Proceedings of the National Conference on Artificial Intelligence*, 247-250.

Maynard Smith and Haigh 1974 Maynard Smith, J. and J. Haigh. The Hitch-hiking Effect of a Favorable Gene. *Genetic Res.*, Cambridge **23** 23-35.

McClelland and Rumelhart 1981 McClelland, J.L. and Rumelhart, D.E. An Interactive Activation Model of Context Effects in Letter Perception. Part 1, An account of basic findings. *Psychological Review*, **88**, 375-407.

McDermott 1980 McDermott, J. R1: an Expert in the Computer Systems Domain. *Proceedings of the National Conference on Artificial Intelligence*, **1**, 269-271.

McDermott 1981 McDermott, J. R1: the Formative Years. *The AI Magazine* **2**, 21-29.

Metropolis *et al* **1953** N. Metropolis, A. Rosenbluth, M. Rosenbluth, A. Teller, and E. Teller. Equations of State Calculations by Fast Computing Machines. *Journal of Chemical Physics* **21**, 1087-1091.

Michalski *et al* **1986** Michalski, R., Carbonell, J., and Mitchell, T. *Machine Learning*, Vol. II, Morgan Kauffman.

Minsky 1986 Minsky, M. *The Society of Mind.* New York, Simon and Schuster.

Minsky and Papert 1969 Minsky, M. and Papert, S. *Perceptrons: An Introduction to Computation Geometry*, MIT Press.

Mitchell *et al* 1983 Mitchell, T. M., Utgoff, P. E., & Banerji, R. Learning by Experimentation: Acquiring and Refining Problem-solving Heuristics. In R. S. Michalski, J. G. Carbonell & T. M. Mitchell (Eds.), *Machine learning, an artificial intelligence approach*, Palo Alto, California: Tioga.

Mozer 1984 Mozer, M. C. *The Perception of Multiple Objects: a Parallel, Distributed Processing Approach.* Unpublished thesis proposal, Institute for Cognitive Science, University of California at San Diego. La Jolla, CA.

Petit and Swigger 1983 Petit, E., and K.M. Swigger. An Analysis of Genetic-based Pattern Tracking and Cognitive-based Component Tracking Models of Adaptation. *Proceedings of the National Conference on Artificial Intelligence* **3**, 327-332.

Reilly 1984 Reilly, R.G. A Connectionist Model of Some Aspects of Anaphor Resolution. *Proceedings of the Tenth International Conference on Computational Linguistics*, Stanford, July, 144-149.

Rendell 1983 Rendell, L.A. (1983). A Doubly Layered, Genetic Penetrance Learning System. *Proceedings of the National Conference on Artificial Intelligence* **3**, 343-347.

Rendell 1985 Rendell, L.A. Genetic Plans and the Probabilistic Learning System: Synthesis and Results. In Grefenstette 1985, 60-73.

Riolo 1986a Riolo, R.L. CFS-C: A Package of Domain Independent Subroutines for Implementing Classifier Systems in Arbitrary, User-Defined Environments. Univ. of Michigan, Div. of Computer Science and Engineering, Logic of Computers Group Technical Report, January.

Riolo 1986b Riolo, R.L. LETSEQ: An Implementation of the CFS-C Classifier System in a Task-Domain that Involves Learning to Predict Letter Sequences. Univ. of Michigan, Div. of Computer Science and Engineering, Logic of Computers Group Technical Report, January.

Rose and Steele 1986 Rose, J. and Steele, G. C* Language Quick Reference. Thinking Machines Technical Report, December.

Rosenblatt 1962 Rosenblatt, F. *Principles of Neurodynamics.* Spartan Books, New York.

Rosmaita 1985 B. J. Rosmaita. *Exodus: An Extension of the Genetic Algorithm to Problems Dealing With Permutations.* M.S. Thesis, Computer Science Department, Vanderbilt University, August.

Rumelhart *et al* 1985 Rumelhart, D.E., Hinton, G.E., and Williams, R.J. Learning Internal Representations by Error Propagation. ICS Report 8506, UC San Diego, September.

Rumelhard and McClelland 1986 Rumelhart, D. E. & McClelland, J. L. (Eds.), *Parallel Distributed Processing: Explorations in the Microstructure of Cognition.* Volume 1. Cambridge, MA: Bradford Books.

Sampson 1976 Sampson, J.R. (1976). *Adaptive Information Processing: An Introductory Survey.* New York: Springer-Verlag.

Schaffer 1984 J. David Schaffer. *Some Experiments in Machine Learning Using Vector Evaluated Genetic Algorithms.* Ph.D. Thesis Dept of Electrical Engineering, Vanderbilt University, December.

Schaffer 1985 Schaffer, J.D. Learning Multiclass Pattern Discrimination. Grefenstette 1985, 74-79.

Schaffer and Grefenstette 1985 Schaeffer, J.D. and J.J.Grefenstette). Multi-objective Learning Via Genetic Algorithms. *Proceedings of the International Joint Conference on Artificial Intelligence* **9**, 593-595.

Sejnowski 1986 Sejnowski, T.J. NETtalk: A Parallel Network that Learns to Read Aloud. Johns Hopkins University Electrical Engineering and Computer Science Technical Report JHU/EECS-86/01, January.

Seknowski *et al* in press Sejnowski, T.J., Keinker, P.K., and Hinton, G.E. Symmetry Groups with Hidden Units: Beyond the Perceptron. *Physica D.*

Selman 1985 Selman, B. Rule-Based Processing in a Connectionist System for Natural Language Understanding. Technical Report CSRI-168, Computer Systems Research Group, University of Toronto, April.

Smith 1985 Smith, D. Bin Packing with Adaptive Search. In Grefenstette 1985, 202-206.

Smith 1980 Smith, S.F. A Learning System Based on Genetic Adaptive Algorithms. PhD Dissertation, University of Pittsburg.

Smith 1983 Smith, S.F. Flexible Learning of Problem Solving Heuristics Through Adaptive Search. *Proceedings of the 8th International Joint Conference on Artificial Intelligence*, 422-425.

Smith 1984 Smith, S.F. Adaptive Learning Systems. In R.Forsyth, (Ed.), *Expert Systems - Principles and Case Studies*, Associated Book Publishers Ltd.

Smith and DeJong 1981 Smith, T.R. and DeJong, K.A. Genetic Algorithms Applied to the Calibration of Information Driven Models of US Migration Patterns. *Proceedings of the 12th Annual Pittsburgh Conference on Modelling and Simulation*, 955-959.

Sontage and Sussman 1985 E. Sontag and H. Sussman. Image Restoration and Segmentation Using the Annealing Algorithm. In *Proceedings of the 24th Conference on Decision and Control*, Ft. Lauderdale, FL, December.

Steele 1984 Steele, G.L. *Common Lisp: The Language*, Digital Press, Massachusetts.

Stefik 1981a Stefik, M. Planning with Constraints (MOLGEN: part 1). *Artificial Intelligence: An International Journal* **16**, 111-140.

Stefik 1981b Stefik, M. Planning and Meta-planning (MOLGEN: part 2). *Artificial Intelligence: An International Journal* **16**, 141-170.

Stein 1977 D. Stein. *Scheduling Dial-a-Ride Transportaion Systems: An Asymptotic Approach.* Ph. D. Dissertation, Harvard University.

TMI 1986 Thinking Machines Corporation, Introduction to Data Level Parallelism. Thinking Machines Technical Report, April.

Touretzky 1986 Touretzky, D. S. BoltzCONS: Reconciling Connectionism with the Recursive Nature of Stacks and Trees. *Proceedings of the Eighth Annual Conference of the Cognitive Science Society*, Amherst, Mass., August, 522-530.

Touretzky 1986 Touretzky, D. S. Representing and Transforming Recursive Objects in a Neural Network, or "Trees *do* Grow on Boltzmann Machines." *Proceedings of the 1986 IEEE International Conference on Systems, Man, and Cybernetics*, Atlanta, GA, October, 12-16.

Touretzky and Hinton 1985 Touretzky, D. S. & Hinton, G. E. Symbols among the Neurons: Details of a Connectionist Inference Architecture. *Proceedings of the Ninth International Joint Conference on Artificial Intelligence*, Los Angeles, CA, August, 238-243.

Touretzky and Hinton 1986 Touretzky, D. S. & Hinton, G. E. A Distributed Connectionist Production System. Technical report CMU-CS-86-172, Computer Science Department, Carnegie Mellon University, Pittsburgh, PA. December.

Touretzky and Derthick 1987 Touretzky, D. S. & Derthick, M. A. Symbol Structures in Connectionist Networks: Five Properties and Two Architectures. To appear in *Proceedings of IEEE COMPCON Spring 1987*, San Francisco, CA, February.

Vecchi and Kirkpatrick 1983 M. Vecchi and S. Kirkpatrick. Global Wiring by Simulated Annealing. *IEEE Transactions on Computer-Aided Design*, Vol CAD-2, No. 4, October, 215-222.

Wagner 1969 Wagner, H.M. *Principles of Operations Research.* Prentice-Hall, New Jersey.

Waltz and Pollack 1985 Waltz, D.L. and Pollack, J.B. Massively Parallel Parsing. *Cognitive Science*, **9**, 51-74.

Waterman 1986 Waterman, D.A. *A Guide to Expert Systems.* Reading, MA: Addison-Wesley.

Wilcox 1985 Wilcox, B. Reflections on Building Two Go Programs. *SIGART Newsletter*, October.

Wilson 1985a Wilson, S.W. Machine Learning of Visual Recognition Using Genetic Algorithms. In Grefenstette 1985, 188-196.

Wilson 1985b Wilson, S.W. Knowledge Growth in an Artificial Animal. in Grefenstette 1985, 16-23.

Wilson 1986a Wilson, S.W. Classifier System Learning of a Boolean Function. Research Memo RIS-27r, Rowland Institute for Science, Cambridge Mass.

Wilson 1986b Wilson, S.W. Classifier Systems and the Animat Problem. Research Memo RIS No. 36r, The Rowland Institute for Science, Cambridge. (To appear in *Machine Learning.*)

Winograd 1983 Winograd, T. *Language as a Cognitive Process.* Reading, MA: Addison-Wesley Publishing Company.

Wright 1977 Wright, Sewall. *Evolution and the Genetics of Populations, Volume 4, Experimental Results and Evolutionary Deductions.* Chicago: University of Chicago Press.